W9-BCT-591

SUPERSTRUCTURES IN SPACE

FROM SATELLITES TO SPACE STATIONS—A GUIDE TO WHAT'S OUT THERE

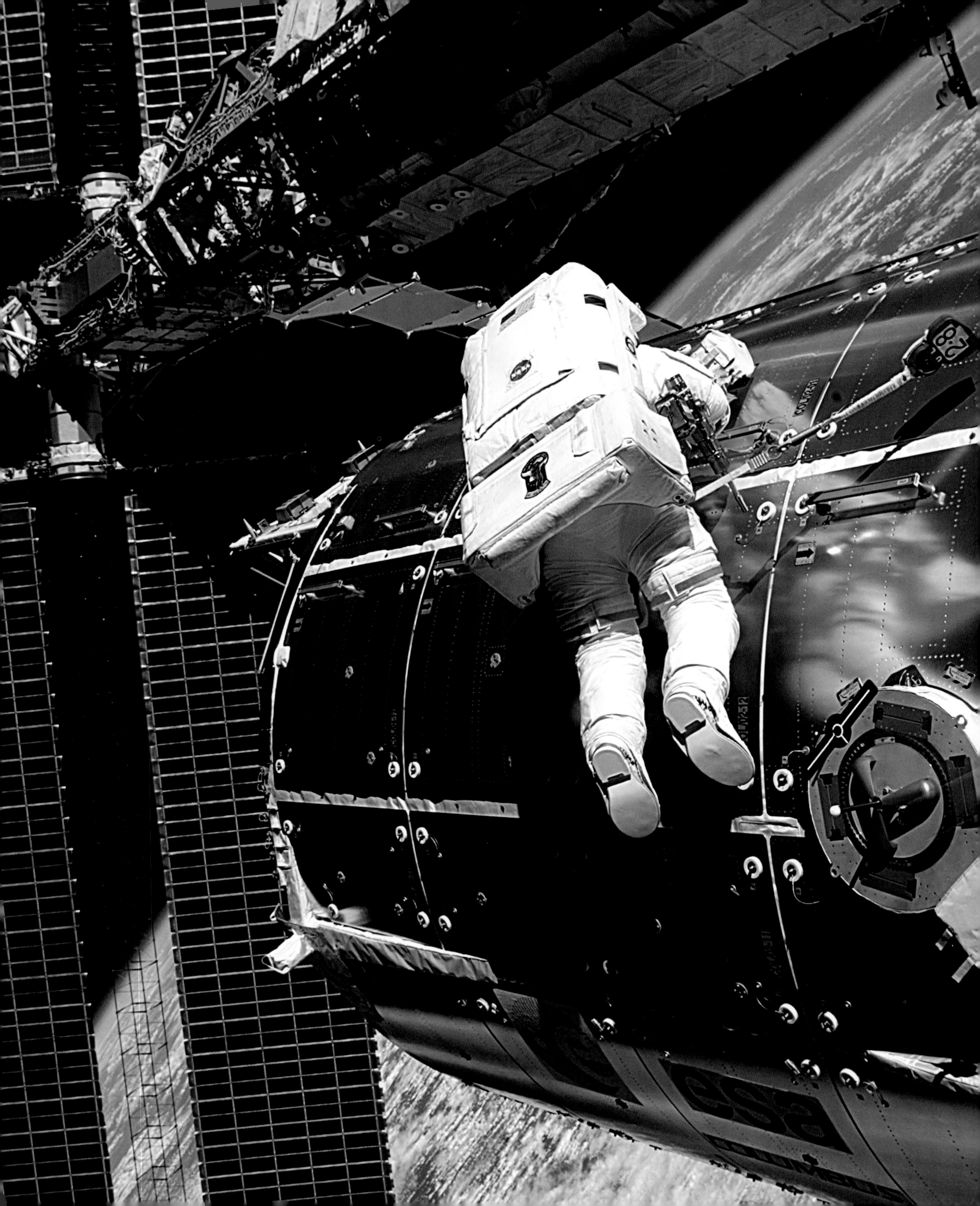

SUPERSTRUCTURES IN SPACE

FROM SATELLITES TO SPACE STATIONS—A GUIDE TO WHAT'S OUT THERE

MICHAEL H. GORN

MERRELL
LONDON • NEW YORK

Introduction
What to Choose?

A ghostly Moon is seemingly suspended in Earth's atmosphere, in a photograph taken in November 2006.

The great dilemma faced by the author of a book of this kind, in which the choices are so vast and the space is so limited, is one of selection. Hundreds of active spacecraft circle the world, fly to other planets, approach the Sun, and encounter comets and asteroids. Thus, the fifty-seven projects featured between these covers constitute only a fraction of the aggregate—but hopefully a representative fraction.

Indeed, the true number of spacecraft flying around the Earth can only be guessed at. Perhaps the best estimate comes from the U.S. Space Command. This combined Navy, Air Force, and Army group operates the Space Surveillance Network (SSN), which consists of twenty-five ground-based radars and optical sensors scattered across the Earth, from Diego Garcia in the Indian Ocean, to Ascension Island in the Southern Atlantic, to the United Kingdom. Space Command tracks objects from about the size of a baseball (4 in./10 cm) and up, and since the launch of *Sputnik 1* in 1957, has found about 24,500 of them in orbit around the Earth. In the first decade of the twenty-first century, Space Command estimates that roughly 8000 entities (both space debris and spacecraft) are circling the world at any one time, of which some 7 percent (560) represent the satellite population. Although this book also covers spacecraft that do not orbit, but leave the gravitational force of the Earth to explore other worlds, for the most part it concerns itself with the orbiting machines, since these are the most common.

The process of selection first involved an assessment of which space programs to feature. The choice was not so much between the enterprises of individual countries as between the four global space programs that exist in the world today. The first of these is *military*, designed to help the armed forces watch for hostile actions from other states, provide precise positioning for military forces, and otherwise lend assistance on the highly technological battlefields of today. The second is *intelligence*, in which nations attempt to glean the capabilities, resources, and intentions of friends and adversaries through intense surveillance. The third is *commercial*, operated by private capital, or government–industry partnerships, with the objective of marketable products of some kind, such as Global Positioning Systems (GPS) for automobiles, trucks, and ships; intensive mapping of urban or agricultural areas; or the fabrication of spacecraft by industry on contract to the military and intelligence sectors.

Each of these three space programs functions behind closed doors, to a greater or lesser degree. Military and intelligence operations generally work in secret, protected by the needs of national security from the public, the media, and scholarly inquiry. Those who sell space-related services secure their discoveries and techniques through legally enforced proprietary protections.

The fourth type of space program bears the stamp of President Dwight D. Eisenhower, and the Cold War in which he governed. At the very time that he and his inner circle of advisers erected the hidden institutions and technologies necessary to pursue spaceflight for military and intelligence purposes during the 1950s, the president also sponsored an open, publicly proclaimed program he called *civil space*. In doing so, Eisenhower and his administration made the conscious decision to separate the American space efforts into two, largely disconnected spheres, one known and the other unknown, a practice very much still in effect in the early twenty-first century, and prevalent throughout the world. Contrary to the other programs, civil space activities glory in publicity and popular attention, and the nations that participate in them seem delighted to present a host of details, partly for national pride and prestige, and partly to position their industries to compete in the global market. As public initiatives, civil space programs

The Space Shuttle *Atlantis* being moved by crawler-transporter to Pad 39A, Kennedy Space Center, Florida.

The launch of *Atlantis* from Pad 39A, Kennedy Space Center, a familiar sight since April 1981. The Shuttle is scheduled for retirement in 2010.

Astronauts attempt to fix a jammed solar array on the International Space Station in December 2006.

have become part and parcel of the political process and the national discourse of every country that engages in them.

SuperStructures in Space concentrates on the world's civil space programs, leaving aside the commercial, military, and intelligence endeavors. (The rare civil space programs that operate out of the public eye are not covered here either.) By defining the scope in this way, the projects under discussion may be presented in a consistent and full, rather than piecemeal, fashion. Moreover, this approach enables *SuperStructures in Space* to reveal not only how these probes, satellites, space stations, and spacecraft operate, but also how they originated and what importance they have in the world at large.

Readers will find fifty-seven spacecraft featured in *SuperStructures in Space*. They reflect a wide variety of countries, many sizes and shapes, a wealth of technologies, and numerous objectives. All of them are seminal in their own way. Each of the four chapters concentrates on a single mission: "Human Spaceflight," "Earth Observation," "Exploring the Solar System," and "Exploring the Universe."

The human spaceflight chapter concentrates on the means by which people are transported into space, and what they do there. The American Space Shuttle is the biggest single carrier of people and payload of any spacecraft. It also symbolizes high cost and a proclivity for failure, realized in two tragic accidents. On the other hand, it constitutes an extraordinarily long technological stride over what came before—a leap unlikely to be equaled by its successors. It is also aging, problematical, and due to be retired in 2010. Perhaps its single most prominent feature—the Canadarm—imparts to the Shuttle its special capacity to transport and release objects as large as railroad cars, to capture spacecraft in need of repair or maintenance, to serve as the main platform during servicing missions, and to fly cargo home from space to precise landing points. For now, the Shuttle–Canadarm combination is mainly being used to continue the assembly of the International Space Station (ISS).

Among both past and present human spaceflight endeavors, the ISS embodies one of the largest and most complicated construction projects ever attempted, whether on the Earth or in space. After a decades-long and tortuous development period in which no one could predict its final profile, it began to take form in the late 1990s. Once the U.S. and the former Soviet Union had decided to collaborate closely, prospects brightened considerably, as American resources and Russian experience with *Mir* combined to rescue the floundering project. After contributing the initial components to the ISS, the Russians and Americans shared the transport of supplies, equipment, structures, and people, and added new parts over time. Meanwhile, the European Space Agency (ESA), the Canadian Space Agency (CSA), the Japanese Aerospace Exploration Agency (JAXA), and others joined the project and developed essential components for the massive structure.

Earth observation, although less exciting than human spaceflight, has evolved into perhaps the most necessary of all civil space endeavors. Arguably the greatest breakthrough related to watching our planet occurred not in the technical developments, but in the conceptual framework. As a result of the publication of the startling pictures of the Earth taken by the Apollo astronauts, as well as the formation of the environmental movement during the 1970s, the Earth came to be viewed as a system, in which key environmental changes had cascading effects rather than local or limited consequences. As a result, during the 1990s NASA's Earth-orbiting system of satellites

A plume of ash rises from the Cleveland Volcano, Alaska, in May 2006, as seen from the ISS.

Comet Tempel 1, sixty-seven seconds after it was struck by the impactor portion of the *Deep Impact* spacecraft.

(among many others) wove the various strands of environmental inquiry into an integrated planetary tapestry. The six satellites of the A-Train, for instance, represent this philosophy of integrated planetary viewing. They fly in very tight formation and make coordinated observations of clouds, aerosols, rainfall, and other aspects of weather prediction, improving the overall accuracy and multiplying the value they might have had singly.

Curiosity, not just practicality, has also played a decisive role in the exploration of space. This explains in part the many probes and satellites sent to various parts of the Solar System from the early days of spacefaring. At such places as the Jet Propulsion Laboratory (JPL) and similar institutions around the world, scientists joined with engineers to develop the spacecraft capable of flying to the far ends of the planetary system, and even beyond it. As the missions progressed through the 1960s, ambitions rose. By the 1970s, researchers not only dreamed about but actually sent robotic vehicles to distant Jupiter, Saturn, Uranus, and Neptune, and finally unleashed them on interstellar space. More recent times have witnessed the Mars Rovers traveling for years on the Red Planet, and the ringed behemoth Saturn and its moons being observed close up by *Cassini* from above, and by its partner, the *Huygens* probe, on the surface of Titan.

Finally, some of the vehicles flying in space exist to answer profound questions about human and celestial origins. Those devoted to comprehending the makeup of the universe attempt to fulfill an engrained human desire to understand the material basis of creation. One of the first and cleverest insights in the pursuit of this enigma had a simplicity that could not be denied: take telescopes out of the soupy atmosphere of the Earth and place them in orbit where they can see without hindrance. This suggestion, heard as early as the late 1940s, evolved during the 1960s, and by the 1990s resulted in gigantic and specialized space telescopes scanning the distant galaxies for gamma-ray bursts here, X-ray sightings there, and ultraviolet events elsewhere. Together, these telescopes have confirmed some theories, raised new questions, and presented an overall portrait of the wild, bizarre, and violent worlds beyond our benign and verdant home planet. It remains to be seen whether the next generation of space observatories—embodied by the James Webb telescope, among others—merely confirms the first generation of galactic sightings, or plunges deeper into the greatest mysteries of all.

ISS and Space Shuttle crews dine together in the Russian module *Zvezda*.

OPPOSITE
This haunting image of the giant Jupiter and its moon Io was sent home by the *New Horizons* spacecraft on its way to far-off Pluto in 2007.

Human Spaceflight
An Uncertain Voyage

1

Introduction

Despite all that has been achieved by human beings in space, people discovered the means to become completely free of the Earth's atmosphere only fifty years ago. On the scale of recorded time, which begins roughly 2500 years ago—with the first true history, written by the Greek chronicler Thucydides—the age of spaceflight represents just two generations of the one hundred that have lived since then. Put differently, if these 2500 years could be compressed into twelve months, men and women would have been spacefarers for only a week.

Because our interlude beyond the Earth has been so brief, no one can say where space travel may lead our species, any more than the Spaniards who washed ashore on Hispaniola with Columbus could have conceived of Mexico City, Rio de Janeiro, or Buenos Aires; or, alternatively, could have guessed that diseases borne by their own bodies would do far more to ravage the indigenous peoples than the sword or rifle.

Given such vagaries as these on our familiar planet, can anyone really predict the future of people in space? Will the physical, physiological, and psychological barriers to long-distance spaceflight—not to mention the expense—persuade the world that limited exploration might be achievable, but that colonization lies beyond reach? Or perhaps once whetted, the appetite for conquest will possess humanity as it did the Spaniards and others after 1492. No one can prophesy reliably.

For now, space travel involving the human element remains a tentative and imperfectly understood venture at best, not just unpredictable, but dangerous. The Columbia Accident Investigation Board (CAIB) Report in 2003 laid bare one of the essential weaknesses of the spacecraft most associated with the world's biggest space program. Rather than an operational vehicle, as its early proponents characterized it, the iconic U.S. Space Shuttle has behaved much more like an experimental aircraft—similar to the X-15 and the other X-planes—than the initial (and optimistic) analogies of a "space truck" and commercial airliner. As the CAIB concluded:

> [The Shuttle] cannot be launched on demand, does not recoup its cost, no longer carries national security payloads, and is not cost-effective enough, nor allowed by law, to carry commercial satellites. Despite efforts to improve its safety, the Shuttle remains a complex and risky system that remains central to U.S. ambitions in space. *Columbia*'s failure to return home is a harsh reminder that the Space Shuttle is a developmental vehicle that operates not in routine flight but in the realm of dangerous exploration.[1]

On the other hand, the problems that have beset the Space Shuttle do not stand alone, but instead reflect the overall level of technical sophistication of human spaceflight at the present time. Indeed, the world's second-largest space program has done no better than the first in providing its spacefarers safe, frequent, and scheduled travel outside the atmosphere. At the time of the *Columbia* accident in February 2003, the Russian Soyuz spacecraft had been flying for thirty-six years, the Space Shuttle for twenty-two. Up to that point, the Shuttle had succeeded in 111 of 113 attempts (launch and recovery), versus 87 in 91 attempts (launch and recovery) for Soyuz. The Russians had suffered two failed launches (both crews saved by abort systems) and two failed re-entries (both crews died). In contrast, the U.S. had experienced one accident on launch and one on re-entry, resulting in the loss of both astronaut crews. In the final analysis, the two nations differed only slightly in their rates of realized reliability: 98 percent for the U.S., 96 percent for the USSR/Russia. Of course, such losses would be intolerable in any other mode of transportation.

Yet, much can be celebrated from the short time aloft: footprints on the Moon; long and trying spacewalks; the on-orbit repair of wickedly complex machines; the construction in space of an immense station; the mastery of techniques to maneuver and dock in space; extensive scientific experiments; and the evident capacity of human physiology to endure (if imperfectly) months of weightlessness.

Paradoxically, although born of intense Cold War rivalries, the feat of sending men and women into space has been pursued over the last decade in an atmosphere of broad international cooperation. In the end, this achievement may signal the greatest success of spaceflight to date: unifying the world so that human beings might leave it.

OPPOSITE
Swedish astronaut Christer Fuglesang takes his third space walk during Space Shuttle mission STS-116 in 2006.

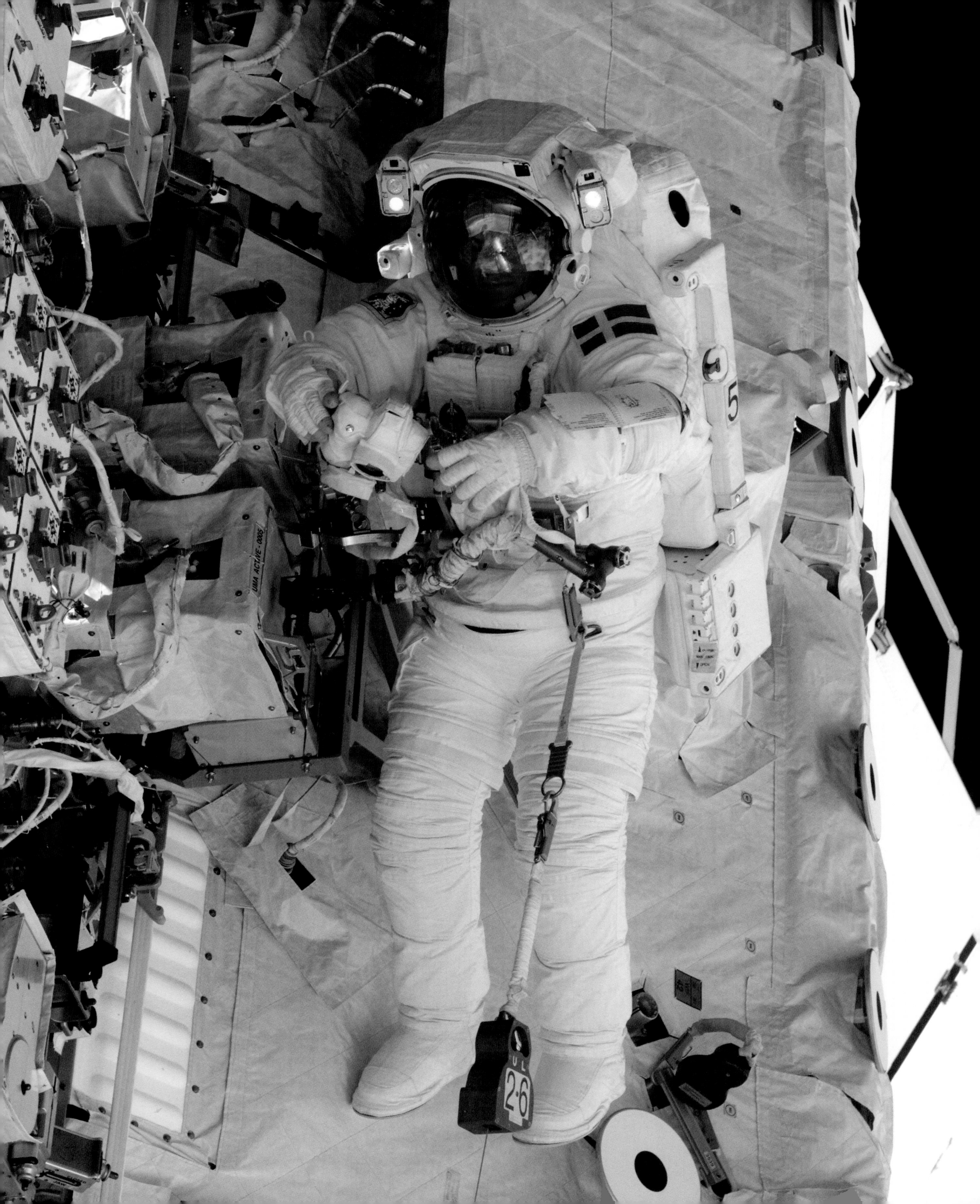
5
2·6

The Space Shuttle

The Orbiter *Atlantis* is lowered by a sling to the external fuel tank and solid rocket boosters arrayed on the mobile launcher platform.

Despite problems of age, complexity, cost, and reliability, the Space Shuttle—even as it nears the end of its service life—remains perhaps the greatest engineering marvel of the early space age. It is not merely the first aircraft to be flown in orbit, but is also as big as an airliner (122 ft/37.2 m in length, with a wingspan of 78 ft/23.8 m), weighs about 250,000 lb (113,400 kg) at launch, has a cargo bay the size of a bus (15 x 60 ft/4.6 x 18.3 m), and is equipped with a manipulator arm as delicate as a tapping finger and as rugged as a claw. More surprising, it came into being in a leap. It supplanted the Mercury, Gemini, and Apollo capsules without evolutionary intermediaries. It replaced limited pilot control with a true cockpit and the full range of piloting options. It replaced cramped seating and very limited cargo space with captain's chairs and enough capacity to send aloft massive girders, complex scientific experiments, and some of the biggest spacecraft ever launched. Finally, before the Shuttle, astronauts splashed down somewhere in an ocean, with all attendant perils. Beginning in 1981, they landed predictably and precisely on specially prepared runways that took them almost to the doorstep of postflight medical and debriefing facilities.

For all of its advancement, however, the Space Shuttle is actually an artifact of the first decade in space. It represents the most conspicuous part of an ambitious post-Apollo space architecture, proposed during the 1960s by human spaceflight advocates inside and outside of NASA. These engineers, scientists, and industrialists envisioned an immense, permanent space station circling Earth as the next great space objective after the Moon, a concept that had gripped the American imagination since the late 1940s. The proponents of the subsequent NASA space station designs conceived of a space "tug" or Shuttle not as an independent entity, but as an integral component of the station, required for its supply and construction, and to ferry passengers to and fro.

These ambitions collided with the decision of President Richard Nixon and his advisers to arrest the expansion of the federal budget. They rejected the overall space station concept, but retained a detail—the tug, or Space Shuttle, itself. Paradoxically, this single vestige of the once mighty station eventually embodied the American space program as nothing else.

The Space Shuttle drew inspiration from a number of sources: from the Air Force's Dyna-Soar project, a boost-glide orbiting weapons system; from NASA's lifting body program, which demonstrated the flying qualities of wingless, high-lift aircraft; and from the famous X-15 rocket plane that reached the edge of space at hypersonic speed and flew home without power. All of them involved aircraft that landed on runways upon return from space. So the Space Shuttle's novelty hinged not so much on its landing as on its take-off. During the late 1960s, engineers at NASA and in industry developed the concept of a "stage-and-a-half" rocket stack, in which reusable solid-propellant tanks would be jettisoned and recovered for reuse while the main module accomplished its mission and flew home. The final design added a gigantic, expendable liquid-fuel tank. North American Rockwell won the competition to build the Orbiter in 1972, and seven years later *Columbia* rolled out of the company's plant.

When the Shuttle finally stood on Pad A at the Kennedy Space Center on April 12, 1981—the day of its maiden flight—nothing like it had ever been seen before. Consisting of four main pieces (two reusable solid rocket boosters, a gigantic external fuel tank, and the Orbiter itself), this new Space Transportation System (STS) aroused intense public curiosity, if only because of its size and difference from the familiar, elongated rockets of the Apollo program. In contrast, the STS on the launch pad looked like a great cathedral (a little like the famed Sacré Cœur in Paris), but with the odd addition of an aircraft mounted vertically on its side. In all, the structure towered 184 ft (56 m) over the Florida landscape and weighed roughly 4.75 million lb (2.15 million kg) with payload and fuel. Upon ignition, with an indescribable roar and issuing a cascade of flame and smoke, the Shuttle's three main rocket engines fired, as did the two solid rocket boosters. Two minutes later, as the STS pulled free of the Earth's atmosphere, the boosters (together

A long view of the Space Shuttle *Discovery* on its launch pad at the Kennedy Space Center, Florida.

Key facts

OVERALL LENGTH:	**184 ft / 56 m**
ORBITER LENGTH:	**122 ft / 37.2 m**
ORBITER WINGSPAN:	**78 ft / 23.8 m**
CARGO BAY:	**15 x 60 ft / 4.6 x 18.3 m**
MANUFACTURER:	**North American Rockwell**
FIRST LAUNCH:	**April 12, 1981**
SCHEDULED RETIREMENT:	**2010**

contributing 5.3 million lb/23.6 million newtons of thrust) fell away, to be recovered and reused. Six-and-a-half minutes later, the empty external tank, which had fed the three main engines, separated and dropped to the Atlantic Ocean below. With two firings of the orbital maneuvering engines, *Columbia* assumed a circular path around the Earth, flying in a Low Earth Orbit (115-400 statute miles/185-644 km on Shuttle flights). During the following two days and thirty-six revolutions, Commander John Young and Pilot Robert Crippen took *Columbia* on a shakedown cruise. Finding all well, they landed at Dryden Flight Research Center at Edwards Air Force Base, California, on April 14. On the ground, however, everyone got an unpleasant surprise. Technicians found that 16 heat-resistant tiles had been lost and 148 damaged. This bad news haunted the Space Shuttle throughout its long service life.

During its more than twenty-five years of service, the Space Shuttle has recorded incredible feats, but none more remarkable than its association with the development, launch, and repair of the mighty space telescope that bears the name of American astronomer Edwin T. Hubble (celebrated, among other things, for experimental verification of Albert Einstein's theory that the universe existed in a state of outward expansion). In fact, a mutual dependency existed among the backers of the *Hubble Space Telescope* (*HST*) and of the Shuttle, since the *HST* gave the Shuttle a heavy-lift customer, and the telescope—the size of a railroad tanker—had no way of getting into orbit by any means other than the Shuttle. One justified the other. But the *HST* seemed star-crossed. The concept of a large optical telescope in orbit had circulated well before NASA existed, originating in the late 1940s and taking the form of design studies during the 1970s. Designs were approved in 1978, but tight budgets in the space agency, cost overruns in the program, and the grounding of the Shuttle from 1986 to 1988 delayed the completion of the telescope for twelve years.

Then, the unimaginable occurred. The telescope was finally launched into orbit in April 1990, but an error in the grinding of its primary mirror resulted in coronas around distant objects. Desperate to rescue its reputation, NASA sent the Shuttle *Endeavour* and a crew of seven back to the *HST* in December 1993 for a daring, in-flight repair mission. The astronauts snatched the telescope, anchored it to the cargo bay, replaced its solar array, installed new corrective optics, and equipped it with an updated camera. The results are known to all. Since then, spacefarers have returned several times to keep *Hubble*'s eye clear to observe the heavens. A final life-extension visit will occur in 2008. (For a fuller treatment of the *HST*, see pp. 144–49.)

Despite its starring role in the *Hubble* drama, the Space Shuttle experienced its own stark failures. On January 28, 1986, Commander Richard Scobee, Pilot Michael Smith, and mission specialists Ellison Onizuka, Christa McAuliffe, Judith Resnik, Gregory Jarvis, and Ronald McNair perished when the Shuttle *Challenger* and the entire Shuttle stack collapsed in an immense explosion a little more than a minute after launch. It resulted from a fiery opening in one of the solid rocket boosters. Seventeen years later, on February 1, 2003, another crew—Commander Rick Husband, Pilot William McCool, and mission specialists David Brown, Laurel Clark, Ilan Ramon, Michael Anderson, and Kalpana Chawla—lost their lives. Seventeen days into their mission aboard *Columbia*, the crew began their descent for Kennedy Space Center, only to die upon re-entry owing to a breach in the Thermal Protection System on the leading edge of the left wing. The *Challenger* investigation (in which Caltech physicist Richard Feynman and the panel's vice chair Neil Armstrong played pivotal roles) and the Columbia Accident Investigation Board (led by Admiral Harold Gehman) both focused on the Shuttle's technical vulnerabilities. But they also acknowledged NASA's impatience with engineers and others who dissented from received wisdom, particularly if their objections threatened launch schedules or inhibited program milestones. As a consequence of these disasters, inherent frailties, and age, NASA officials designated 2010 as the final year of Shuttle operations.

BELOW
***Discovery*'s launch in August 2005 returned the Space Shuttle to operation after a stand down of almost two-and-a-half years.**

BOTTOM
An ISS Expedition 13 crewmember recorded this close-up image of the tail section and main engines of the Orbiter *Discovery*.

A little distorted by a fish-eye lens, this image shows *Atlantis* during mission-115, as it rises from Pad 39B at Kennedy Space Center.

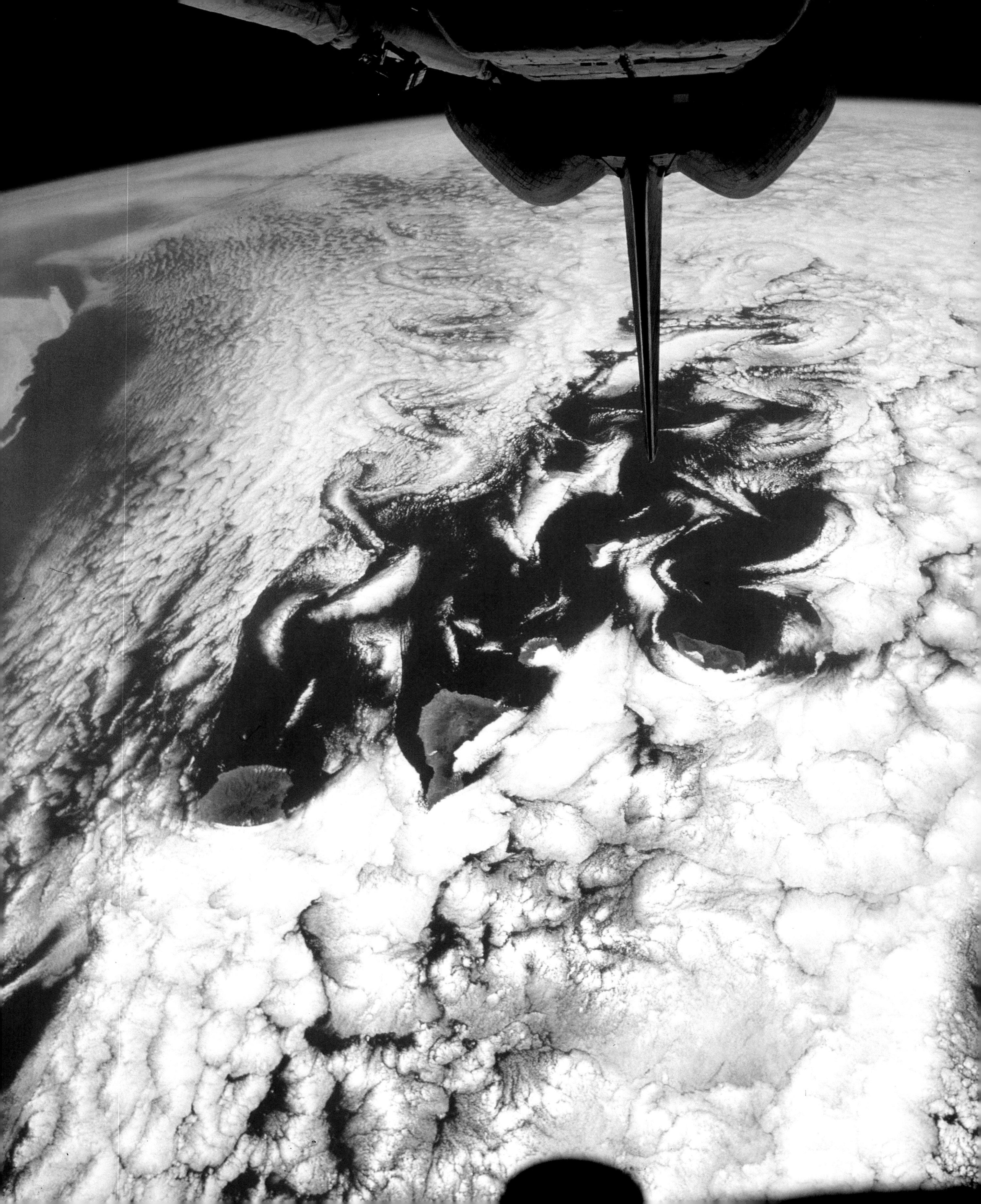

The return to flight mission of the Space Shuttle *Discovery* (July 26 to August 9, 2005) after a long, two-and-a-half-year pause in activity, brought a much-needed sense of renewal to the program. In its final incarnation before retirement, the venerable spacecraft returned to service as an international ambassador. Although the Space Shuttle originated as the signature American spacecraft in the post-Apollo age, in practice it has evolved into the embodiment of global cooperation in space. The construction of the International Space Station beginning in 1998 allowed the Shuttle to fulfill one of its most notable multinational roles: as the only vehicle capable of hauling the mammoth trusses and components to the ISS for assembly. But this cooperative feat paralleled many that had come before. In one of many scientific endeavors, the Space Laboratory, or *Spacelab*, of the European Space Agency (ESA) flew in the cargo bay of eleven Shuttle missions, collecting unique data related to physics, astronomy, life science, and Earth sensing. In the run-up to the initial construction and manning of the ISS, the U.S. and Russia collaborated on a number of Shuttle flights so that they might become mutually familiar with their equipment and procedures. The Shuttle docked with the *Mir* space station nine times between June 1995 and June 1998, transporting astronauts, cosmonauts, food, supplies, and scientific equipment to the Russian spacecraft, where individuals from the two nations shared responsibilities for months at a time. During launches between 1995 and 2000, an intense period of preparation for the ISS, the Shuttle carried thirteen cosmonauts as crewmembers.

Inside its cabins, the Shuttle also flew more than thirty "international astronauts" (defined as individuals who have temporarily joined the American astronaut corps and have trained at Johnson Space Center). They represented the European, Japanese, Brazilian, Canadian, and French space agencies. Many foreign nationals not associated with other space entities also flew, such as Israeli Colonel Ilan Ramon, who perished aboard *Columbia*.

Finally, the Space Shuttle served as an ambassador of sorts for American citizens, too. Because the Shuttle's cabin capacity exceeded that of any other vehicle to date, the Orbiters—perhaps unintentionally—broadened greatly the cross-section of humanity who had been launched into space. During its first twenty-four missions alone, the STS conveyed 125 astronauts, in effect democratizing spaceflight by enabling American women and minorities to join the elite club of space travelers as never before, and, eventually, to fly routinely. By 2007, more than 600 crew had traveled aboard.

In its totality, the U.S. Space Transportation System—of which the Space Shuttle Orbiter is the most recognized part—continues to be the most ambitious creation of the age of spaceflight, technologically superior to all other space vehicles transporting human cargo, past or present. It represents an astonishing leap in sophistication, supplanting in one blow capsules splashing down in the seas with aircraft making pinpoint landings on runways. To this day, the Space Shuttle continues to be a hare among tortoises.

But like the fable from the animal kingdom, there is something to be said for the tortoises. The Shuttle is complex and prone to malfunction; it is expensive and labor-intensive; and it has suffered two complete failures, resulting in the deaths of two entire crews. Nonetheless, it is still a wonder.

OPPOSITE
An image of part of the Canary Islands from space, bisected by the Orbiter *Atlantis*'s vertical stabilizer.

The Orbiter *Atlantis* soars above a mountainous coastline, just after undocking from the ISS during Expedition 1.

A ground-level view of Kennedy Pad 39B, on which *Atlantis* and the rest of the Shuttle stack stand glowing in the morning light.

The Canadarm

Perhaps the Space Shuttle's most heralded claim to an international pedigree is emblazoned on the Orbiter itself. Here, the name "Canada" and the familiar maple leaf signal a partnership that has enabled the Shuttle to undertake its most complex assignments. The first incarnation of Canadian participation dates as far back as 1969, when NASA presented potential foreign partners with a variety of Shuttle system concepts to which they could make contributions. Five years later, Canadian Space Agency (CSA) officials announced their choice: they would research and develop the Space Shuttle Attached Remote Manipulator System (SMRS), later known colloquially as the Canadarm. Project manager Dr. Garry Lindberg interested three firms in collaborating on the design and testing: Spar Aerospace, CAE Electronics, and DSMA Atcon. Canada's National Research Council invested $108 million toward the development of the initial flight hardware.

The Canadarm presented formidable challenges to the scientists, engineers, and technicians who attempted to make it a reality. No off-the-shelf parts or equipment existed for a piece of machinery that could do what NASA wanted: to move deftly and surely, to operate in the unforgiving conditions of space, and to have sufficient reach to grapple objects inside the Shuttle's cargo bay. If these demands failed to impress, the Canadarm also had to be rugged, lightweight, reliable, and safe. In addition, the team encountered a pivotal problem related to testing the arm. Designed to function in near-weightless conditions, it could not lift itself off the ground in the Earth's gravity.

In November 1981—seven years after their country accepted the challenge—Canadians held their breath as the Canadarm unfurled itself from the cargo bay of STS-2 (only the second Shuttle flight) for its first deployment. Commander Richard Truly then gave Canadarm's well-wishers the news they were hoping for: "The arm is out and it works beautifully. Its movements are much more flexible than they appeared during training simulations."[2] Canadarm has flown on most of the Shuttle missions ever since.

In conceptualizing the Canadarm, its designers worked out a simple but brilliant premise. They decided to fashion the machine on a proven model—on the human arm itself. They even patterned its parts on human anatomy: the shoulder joint, the upper-arm boom, the elbow joint, the lower-arm boom, and the wrist joint. (The wrist, elbow, and shoulder joints had three, one, and two degrees of motion, respectively.) Only the hand (called an "effector") bears a non-anatomical name, and it consists of a cylindrical fixture with a wire-snare device that fits over a special prong. These wires envelop the prong and hold on to whatever the arm must raise or move. Furthermore, the arm's actuating parts have analogs in the human arm: nerves made of copper wiring, bones made of graphite fiber-reinforced tubing, and muscles composed of electric motors. The Canadarm's engineers also settled the dilemma of testing a structure too heavy for gravity by constructing a support rig that enabled the machine to flex its joints naturally. The team also relied on computer simulations.

Astronauts who train on the SRMS simulator get the feel of the Canadarm, which, according to Canadian astronaut Mark Garneau, "really does function like an extension of your own body once you become familiar with it."[3] On the flight deck, they use two hand controllers that manipulate the arm one joint at a time, or all six joints in coordination. The translational hand controller moves the machine right, left, down, up, backward, and forward; the rotational controller

The Canadarm is poised for action as it towers over the open Shuttle cargo bay.

Key facts

GO-AHEAD BY CANADIAN SPACE AGENCY:	**1974**
FIRST USE IN SPACE:	**November 1981**
LENGTH:	**50 ft / 15 m**
DIAMETER:	**15 in. / 38 cm**
WEIGHT:	**905 lb / 410 kg**

The Canadarm appears to hang in space without support, suspended over the broad expanse of the Earth below.

PAGES 24–25
Astronaut Michael Gernhardt grasps the end effector of the Canadarm during an Extravehicular Activity (EVA) on STS-104. Over his shoulder is a faint, arrow-shaped outline of the island of Cyprus.

operates yaw, roll, and pitch. Two closed-circuit television cameras help the astronauts get the Canadarm into alignment to do its tasks. As might be expected, the long appendage moves slowly—at the rate of 2 ft (0.6 m) per second empty and 2½ in. (6 cm) per second with the arm loaded.

During actual missions, the arm—905 lb (410 kg) in weight, 50 ft (15 m) in length, and 15 in. (38 cm) in diameter—is a marvel of versatility. It can handle with equal dexterity the capture and anchor of payloads ranging from the very small to the gigantic *Hubble Space Telescope*, and can grasp and lift any of the massive trusses of the International Space Station. For Canadians, the Shuttle arm constitutes not just a novel and remarkable instrument, but also a proud national achievement, brought to mind every time the cargo bay opens and the symbol of the red maple leaf rises against the exquisite backdrop of darkness. But this represents more than just a parochial feeling: just as they did when *Sputnik 1* first orbited the Earth on October 4, 1957, achievements in space continue to confer on the nations that engage in them a special status, probably not to be gained in any other way.

The International Space Station

The concept of massive domiciles in space, where human beings might live and work high above the safety of the Earth, captivated space enthusiasts for much of the twentieth century. Some regarded them as indispensable spaceports, from which future voyages of discovery would embark to distant places, just as mariners once took to the seas in search of adventure and wealth. Others thought of these outposts somewhat differently, more like fortresses on the frontier. Those who accepted this analogy felt that just as the U.S. Army built fortifications to protect and expand settlement in the West, those brave enough to venture into space required strong mustering points from which they could prepare to venture to other worlds.

During the years just after the Second World War, an unlikely coterie of proponents—led by individuals as diverse as rocket scientist Wernher von Braun, artist Chesley Bonestell, amusement-park developer Walt Disney, and author Willy Ley—publicized the objective of a continuous human presence outside the atmosphere. Ley wrote as early as 1944 that space travel depended on space stations. Von Braun's dream, expressed in an essay of 1952 that actually made reference in its title ("Crossing the Frontier") to wilderness, reached its full public expression not in words, but in pictures. Artist Chesley Bonestell's dramatic renderings of orbiting space ships appeared in *Collier's Magazine* in the same year as the publication of "Crossing the Frontier." The majestic, cold beauty and gigantic proportions of Bonestell's hub, spoke, and wheel design captured the American imagination.

Although these images lingered even in the minds of the practical engineers who joined NASA in 1958, it was not until 1969 that space-station development began to be considered seriously. But the real push occurred almost thirty years after the *Collier's* illustrations. With the first flight of the Space Shuttle—a vehicle uniquely capable of hoisting the massive constituents of a station aloft—advocates for a space station began to beat the drum. NASA Administrator James Beggs presented the project to President Ronald Reagan, who approved it despite its high cost. The president, in turn, unveiled it to Congress in 1984 in his State of the Union address as a scientific platform, as well as a platform for international cooperation. Reagan called for a ten-year deadline on completion, more than a little reminiscent of President Kennedy's Apollo declaration twenty-three years earlier.

The concept drew on previous American experiences. The first U.S. incarnation of a space station occurred with the *Skylab* Orbital Workshop, a 200,000-lb (90,700-kg) adaptation of an upper stage of a leftover Saturn V rocket; another Saturn V launched it into orbit in spring 1973. *Skylab* flew unoccupied for two weeks before a three-astronaut crew arrived in an Apollo capsule, launched by a Saturn 1B rocket. Among other experimental equipment, *Skylab* carried the Apollo Telescope Mount, a full-scale solar observatory. More important than the data collected, the three missions to *Skylab* (occurring intermittently between May 1973 and February 1974) gave three astronaut crews a total of 171 days in orbit—a real taste of long-duration flight.

Close on the heels of *Skylab*, NASA collaborated with the Soviet space program on the Apollo–Soyuz Test Project (ASTP), the first space collaboration between superpowers and one that presaged the close relationship that eventually developed between the two leading nations pursuing human spaceflight. Flying in patterns more common to the twenty-first century, Soviets Aleksey Leonov and Valeriy Kubasov left for the on-orbit rendezvous from the Baikonur Cosmodrome in Soviet Kazakhstan, while Americans Vance Brand, Thomas Stafford, and Donald Slayton approached from the Kennedy Space Center. In an act rich with symbolism, the two crews arrived at their meeting on July 17, 1975, aboard the two old space-race rivals—Soyuz and Apollo—and docked. They then migrated freely between the spacecraft, visiting each other four times and conducting experiments before going their separate ways the next day.

Ten years after ASTP, James Beggs introduced the nation to space station *Freedom* in 1985 It looked ungainly and asymmetric, which may have hurt its chances. Unlike the stately, elegant wheel still uppermost in the minds of those who remembered the image propagated in *Collier's*,

OPPOSITE
***Atlantis* undocks from *Mir-19*, bringing home three cosmonauts from the *Mir-18* mission.**

A powerful symbol of the new age of spaceflight, the Shuttle *Atlantis* and the *Mir-19* space station are united in July 1995, photographed by cosmonauts in a Soyuz-TM spacecraft that had detached from *Mir* for this occasion.

Key facts

U.S. SPACE STATION ANNOUNCEMENT:	**1984**
SPACE STATION FREEDOM UNVEILED:	**1985**
INITIAL ESTIMATED COST OF FREEDOM:	**$8 billion (shared by U.S., Japan, Canada, ESA)**
INITIAL ESTIMATED COST OF ISS:	**$17.4 billion (U.S. portion only)**
MATING OF THE FIRST TWO SECTIONS OF ISS:	**December 1998**
SCHEDULED COMPLETION OF ISS:	**2010**

and later in the film *2001: A Space Odyssey*, *Freedom*'s awkward profile incorporated a boom for solar power, twin vertical keels for servicing materials and laboratory equipment, and living quarters and laboratories clustered at the center. Although big—500 ft (152.4 m) in length and 360 ft (109.7 m) in height—it accommodated no more than eight astronauts in a small cabin. Beggs proposed it to Congress as a multi-purpose endeavor, combining an exploration platform, laboratory, observation post, microgravity facility, and satellite-repair shop. Beggs also won the support of the Canadian, Japanese, and European space agencies to share the estimated $8 billion cost. But budget-cutting caught up with *Freedom* almost from the start, and in 1986 all fabrication capabilities disappeared from the blueprints, leaving only a laboratory. The contraction continued, owing in part to the Shuttle flight hiatus caused by the *Challenger* disaster, until only four crewmembers remained, confined to microgravity and life-science experiments.

BELOW
The ISS receding into the distance, as seen from the Space Shuttle *Atlantis*.

BOTTOM
A view of the ISS from the Space Shuttle *Discovery*, taken in December 2006 during undocking maneuvers.

At first, the Clinton administration cut even more, renaming the rump space station *Alpha*. Then NASA Administrator Daniel Goldin persuaded a Russian government eager to save money and divorce itself from the Soviet past to collaborate on the project. This milestone event in the history of spaceflight revived the space station and brought about a fresh energy. Engineers from the two countries substantially redesigned and somewhat increased the size of the crew cabin (allowing for six passengers rather than *Alpha*'s four), added to the length of the central corridor, boosted electrical power generation, and reintroduced the node modules and the U.S. laboratory. Indeed, the new configuration envisioned a final structure weighing twice that of the station *Freedom*. The project also profited from the Russians' long experience with the Salyut and *Mir* spacecraft and their promise to build new modules for this endeavor. Finally, the countries attracted earlier by Beggs joined with the Russians and the Americans anew to form a multinational consortium. They renamed their mammoth undertaking the International Space Station.

But the ballad of the ISS did not end there. Congress at first reacted negatively to the partnership with the Russians, voting in 1993 to cancel funding for the entire project (then estimated at $17.4 billion), a measure that lost in the House of Representatives by just one vote. Two years later, however, the mood improved as the value of the space partnership became evident. Congress approved an annual ceiling of $2.1 billion. Yet even this measure failed to bring stability. It seems that, during 1997 and 1998, NASA went over the $2.1 billion budget limit. Worse still, an independent review predicted that the $17.4 billion maximum might be exceeded by more than $7.3 billion once the ISS had been completed. Moreover, the auditors found that delays ranging from ten to thirty-six months loomed on key ISS components, resulting in a likely end date of 2006 rather than the planned 2004. In fact, 2006 proved to be sanguine. Owing in part to the two-and-a-half-year hiatus in Shuttle launches following the *Columbia* disaster (February 2003 to July 2005), the leaders of all of the space agencies involved in the ISS agreed in March 2006 to make 2010 the final year of construction.

Wisely, the Russians and Americans pooled their efforts long before mating the first parts of the ISS. Engineers and scientists from the U.S. had perhaps the most to gain. The first Soviet space station, *Salyut 1*, although a failure, inaugurated a series of successful spacecraft flown between 1971 and 1986, culminating in *Salyut 7*. From the knowledge gleaned from Salyut, the USSR developed the next-generation space station, *Mir*, a structure weighing more than 200,000 lb (90,700 kg), first launched in 1987. When the two superpowers struck the deal for the ISS in 1993, the Russians had already accumulated twenty-two years of experience with long-duration spaceflight and hoped to continue with *Mir-2*, an unfunded design with little prospect of fulfillment in post-Cold War Russia. Then Dan Goldin and NASA approached. The subsequent ISS represented a merging of *Alpha* and *Mir-2*.

Because *Mir* continued to fly while the ISS rose from the drafting tables, seven astronauts received their first exposure to prolonged periods of spaceflight—up to six months—aboard the aging spacecraft between 1995 and 1998 (during which time two emergencies occurred, one a hair-raising onboard fire, the other a collision with an unoccupied Progress spacecraft). Meanwhile, seven Shuttle missions hosted seven cosmonauts in the same period—a useful step, since the American spacecraft served as an integral part of the ISS architecture.

An artist's impression shows the ISS as it might look upon its completion in 2010.

The serenity of spaceflight: an orbital sunrise viewed from the ISS.

Zarya

Japanese astronaut Koichi Wakata floats through a laden *Zarya*, less than two weeks before permanent occupancy of the ISS began.

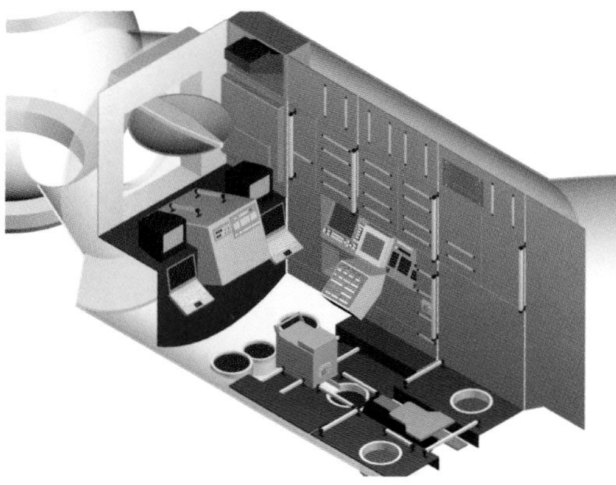

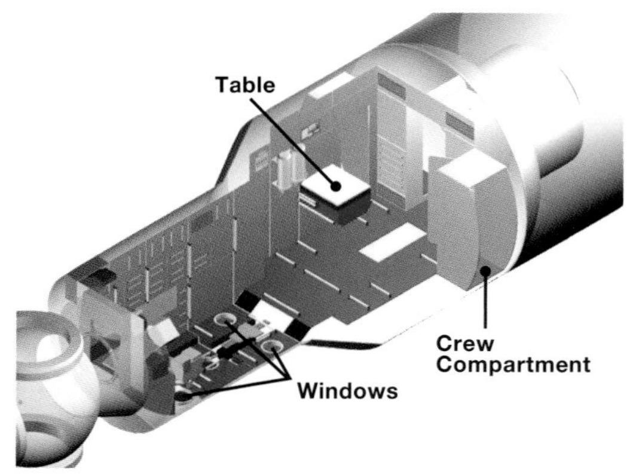

Two cutaway illustrations of the interior of the *Zarya* Control Module.

At last, late in 1998, the ISS's first two components waited to be joined in orbit, almost fifteen years after President Reagan's space station announcement. First came the Russian part, called the *Zarya* Control Module (*zarya* means "sunrise" or "dawn"), a name suggesting a new age of partnership in space. Also known as the Functional Cargo Block (or FGB, following its Russian spelling), this initial part of the ISS owed its origins to the Russian Transport Logistics Spacecraft (TKS), conceived as a ferry for the Almaz military space station. Under agreement with the Boeing Company, the ISS's prime contractor, the Khrunichev State Research and Production Space Center in Moscow received $220 million to construct *Zarya*, and did so between December 1994 and January 1998, within budget and on time. (For the same price, Khrunichev also fabricated *Zarya*'s twin, FGB-2, ready for a contingency.) Lockheed had made a competing bid of $450 million.

The pressurized *Zarya* constituted a complex piece of equipment, more than 41 ft (12.5 m) long, more than 13 ft (4 m) wide, and weighing 42,600 lb (19,300 kg). It had several purposes, not least of which was to be a conduit between American and Russian components. Its three docking ports (at its ends and on one side) would accommodate an American module at one end, a Russian module at the other, and visits from the Progress or Soyuz transport vehicles at the third port. In addition, its big solar arrays (approximately 35 x 11 ft/10.6 x 3.4 m) plus six batteries generated 3 kW of electricity: enough power for itself and the first U.S. ISS component. *Zarya* equipped the ISS with generous fuel storage and with essential maneuvering capability, propulsion, and guidance. Its sixteen external tanks carry 13,200 lb (6000 kg) of propellant, and the spacecraft hauls two large engines, twelve small ones, and twenty-four large steering jets for re-boost and major changes in orbit. Its interior also offers pressurized storage space.

Zarya took off from the Baikonur Cosmodrome and entered the Earth's orbit on November 20, 1998, on the back of a three-stage Russian Proton rocket, a heavy-launch vehicle that has flown more than 200 times, with 98 percent reliability. It began in an elliptical path as high as 211 miles (340 km), and as low as 137 miles (220 km), where it underwent a series of automated commands that brought its systems to life, opened the solar arrays, awakened the communications antennae, and launched a series of operational tests. Its manufacturer expected a service life of at least fifteen years. In succeeding days, its two large engines fired and boosted the spacecraft to a circular orbit of 240 miles (386 km). There, it awaited the arrival of its American mate.

Zarya comes into range of STS-88 prior to mating with the U.S. module _Unity_.

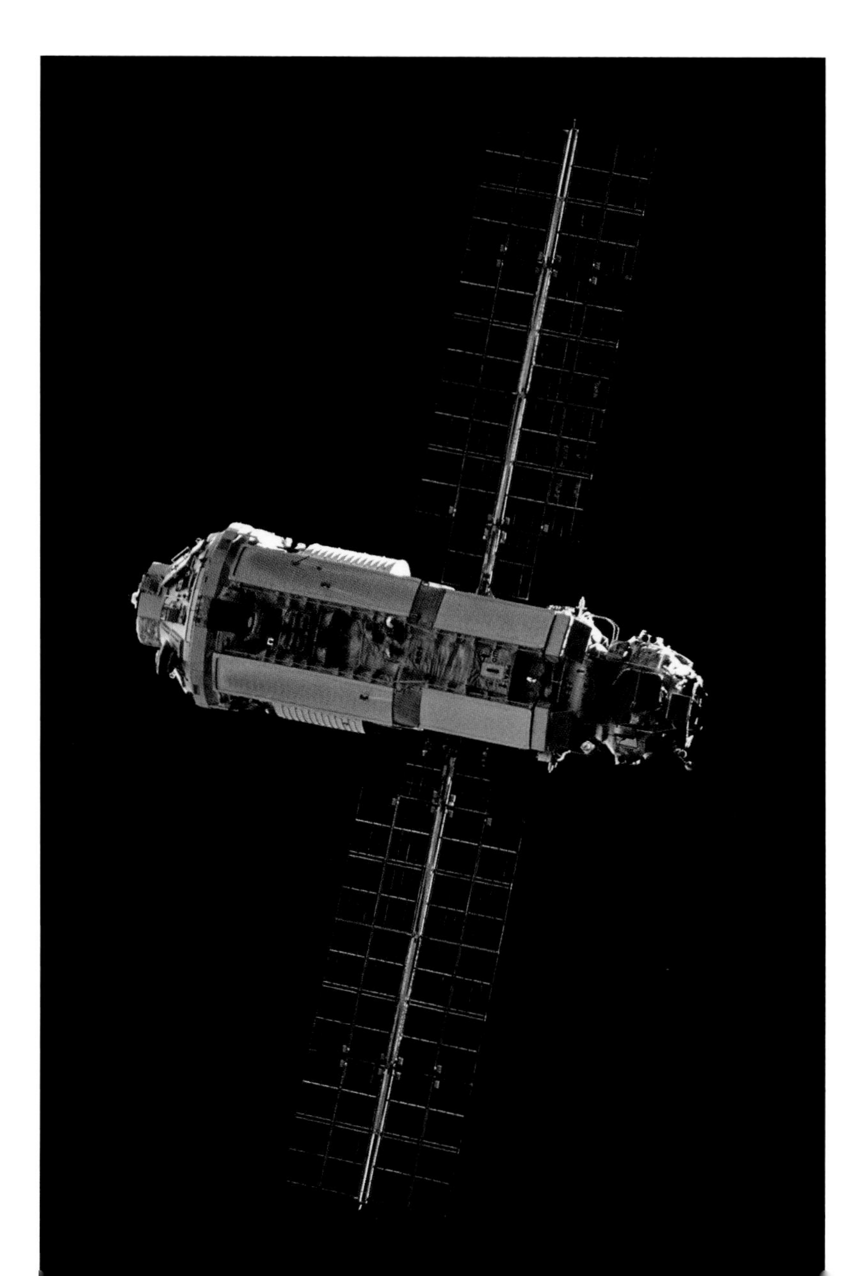

Key facts

MANUFACTURER:	**Khrunichev State Research and Production Space Center**
PERIOD OF CONSTRUCTION:	**1994–98**
COST:	**$220 million**
LENGTH:	**41 ft / 12.5 m**
WIDTH:	**13 ft / 4 m**
WEIGHT:	**42,600 lb / 19,300 kg**
LAUNCH:	**November 20, 1998**

Unity

TOP
The Canadarm grasps *Zarya* and guides it toward the anchored *Unity* module.

ABOVE
The first two parts of the ISS are joined: *Zarya* (top) meets *Unity* in *Endeavour*'s cargo bay, December 1998.

Two weeks after *Zarya*'s launch, on December 4, 1998, Space Shuttle *Endeavour* left Kennedy Space Center in Florida on mission STS-88, the first dedicated to the ISS. In its cabin, one Russian and five Americans anticipated the beginning of an epoch in spaceflight; in its cargo bay, the American-made *Unity* component represented the new era. On this flight, the crew would assemble the initial pieces of the world's first multinational domicile in space, the foundation of one of the biggest construction projects in human history, all achieved outside of the atmosphere.

The crew of *Endeavour* spent the rest of the first day aloft firing the Shuttle's engines to close the distance between themselves and *Zarya*, which was then on its 222nd rotation around the Earth. During the next day (December 5) astronaut Nancy Currie practiced with the Canadarm while the others continued to chase *Zarya* and checked the tools and equipment needed for the capture of the Russian spacecraft. At last, on December 6, Commander Robert Cabana fired one of *Endeavour*'s engines in order to position the Shuttle 600 ft (182 m) below the target. Then, with the cargo bay open and the three-story-high *Unity* standing upright, Cabana took manual control of *Endeavour* and swung the ship on a semicircular arc, starting below *Zarya*, to 350 ft (107 m) in front of it, to 250 ft (76 m) directly above it. As Cabana then lowered the Shuttle toward the objective, Currie waited for the point at which the Russian structure hovered 10 ft (3 m) from the edge of the cargo bay. At that moment, she activated the Canadarm, snared *Zarya*, and drew it into alignment with *Unity*'s docking mechanism. While success rested on the astronaut's skill, it also depended on the arm, which up until this moment had not moved anything as heavy as *Zarya*, the closest previous load having been 6000 lb (2700 kg) lighter. Down to inches, there was some difficulty aligning the twenty-four pins that mated the two spacecraft. But Currie at last lined up the pins and holes, and, having done so, let the Canadarm relax. The commander then fired the Shuttle's thrusters, locking the two spacecraft into place.

The great drama over, the crew set about the sweat work of linking up the electrical and other systems of *Zarya* and *Unity*, procedures entailing three spacewalks (totaling more than twenty-one hours) over several days. This segment of the voyage began the day after the mating, as power flowed into *Unity*. A week after the mission had begun, Commander Cabana led his crew into the American part of the ISS, where, after installing lights, fans, and the S-band communications system, they ventured into *Zarya*. The next day, they brought in tools, clothing, and other goods needed for the crew that would later inhabit the ISS, sealed the hatches, and withdrew to *Endeavour*. On Sunday, December 13, Pilot Rick Sturckow undocked from the *Unity* portion of the International Space Station, pulled back to a position 450 ft (137 m) above it, and turned for home, leaving the ISS to fly unpiloted until the next Shuttle visit.

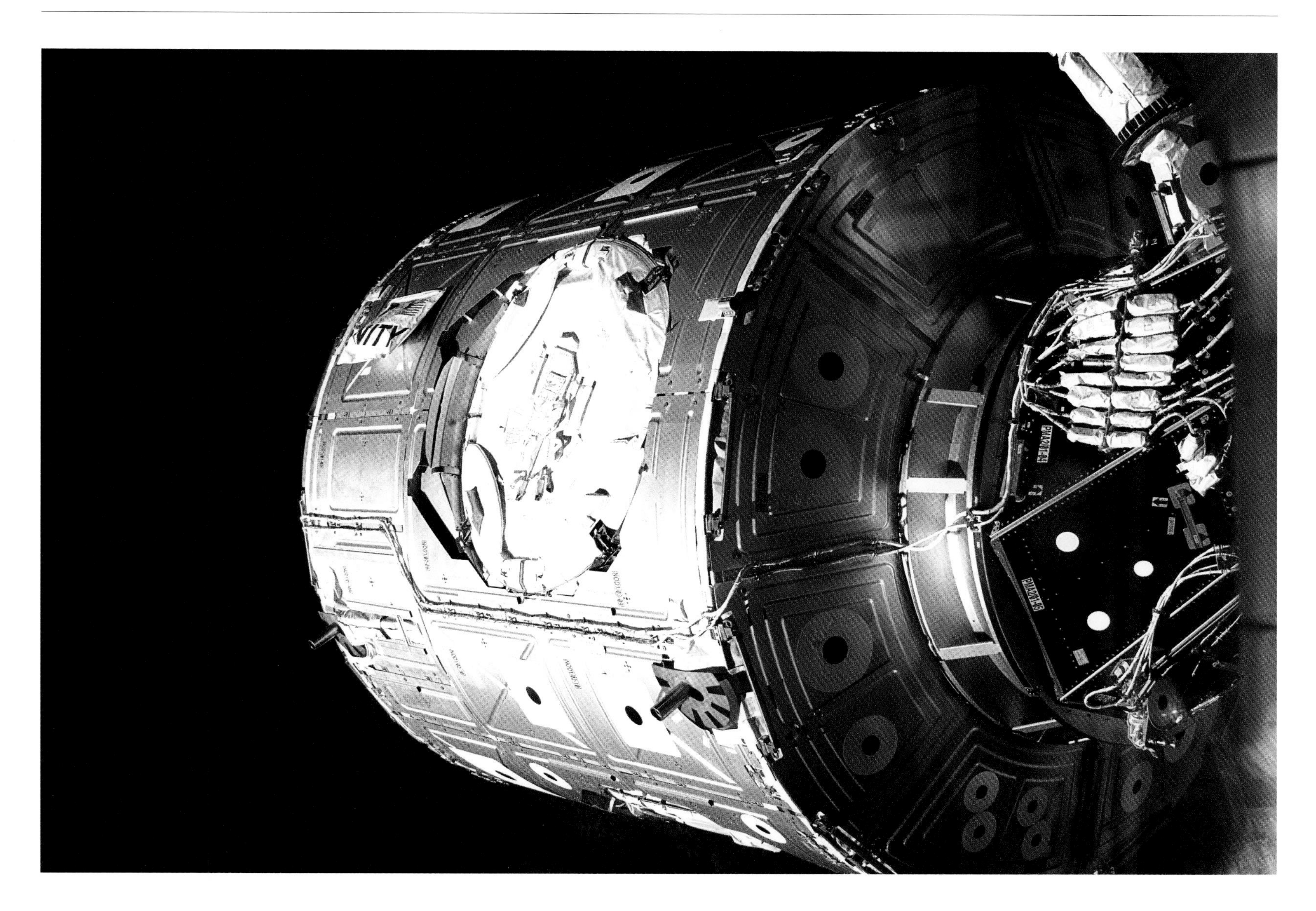

A close-up of the American *Unity* module.

Key facts

MANUFACTURER:	**Boeing Company**
OVERALL STRUCTURE:	**Aluminum**
NUMBER OF MECHANICAL PARTS:	**50,000**
LENGTH:	**18 ft / 5.4 m**
DIAMETER:	**15 ft / 4.6m**
WEIGHT:	**25,600 lb / 11,600 kg**
LAUNCH:	**December 4, 1998**

As the ISS receded from view the astronauts and cosmonaut could be satisfied with their achievement. To *Zarya* they had added *Unity*, also called Node 1, the initial juncture box of the ISS. Fashioned out of aluminum by the Boeing Company at Marshall Space Flight Center in Huntsville, Alabama, the stubby, six-sided *Unity* measured 18 ft (5.4 m) in length, 15 ft (4.6 m) in diameter, and weighed 25,600 lb (11,600 kg). (At its launch aboard *Endeavour* it actually measured double its normal length, extended by long adapters plugged in to connect it to *Zarya* at one end and to the Shuttle at the other.) A port on each of its facets would eventually connect it to six other station components, in turn linking them all to one another through *Unity* itself. To accomplish this objective, it held more than 50,000 mechanical parts, 216 lines dedicated to gases and liquids, and 121 electrical cables measuring 6 miles (9.6 km) end to end. One day, these items would allow *Unity* to join the American laboratory module, the European Space Agency (ESA) Node 3, a truss for the station, an airlock, and a cupola to itself and to *Zarya*.

Zvezda

ABOVE
Cosmonaut Sergei Krikalev, Expedition 1 Flight Engineer, at the porthole on *Zvezda* as the Shuttle *Atlantis* approaches for docking.

LEFT
***Zvezda* under construction at the Khrunichev State Research and Production Space Center in Moscow, Russia.**

BELOW, LEFT
Diners at the *Zvezda* café.

BELOW
Cosmonaut Sergei Krikalev retrieves tools from the *Zvezda* module.

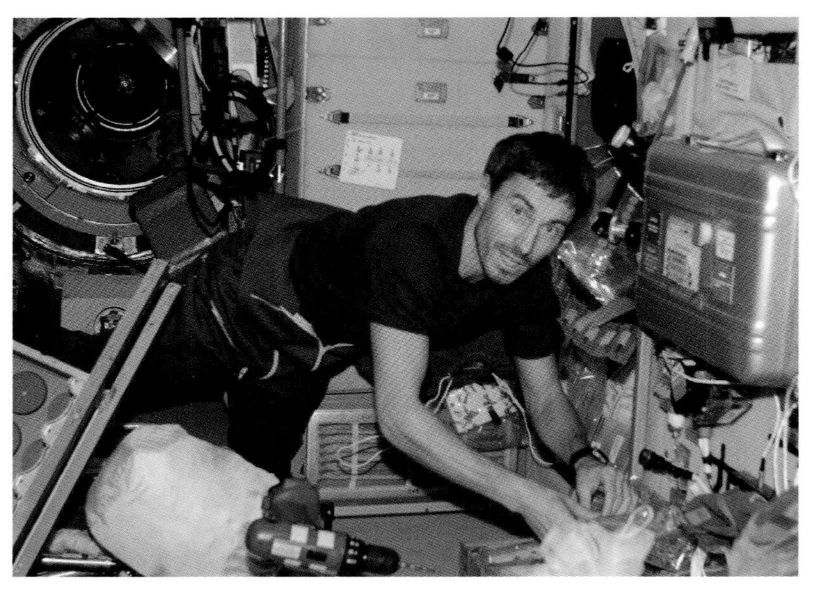

Originally, planners of the ISS expected the third segment to be attached just six months after *Zarya* and *Unity* linked up. As it turned out, this module did not arrive until a year after the mid-1999 target. In between, however, there had been two logistics missions to the ISS, one in May 1999, the other in May 2000, each flight carrying 2,000 lb (907 kg) of supplies, logistics, and equipment. At last, the much-anticipated segment arrived. Called *Zvezda* (meaning "star"), it constituted the first all-Russian designed and fabricated component of the ISS. Patterned on the core section of the venerable *Mir* space station, the 42,000-lb (19,000-kg) *Zvezda* left the Baikonur Cosmodrome on July 12, 2000, on the back of a Proton rocket.

Once in orbit, *Zvezda* underwent a long and impressive series of maneuvers, both automated and originating from Russian ground-control, that aligned it with the *Zarya–Unity* pairing. First, a series of pre-programmed computer commands awakened the Kurs rendezvous and the Lira communications systems, after which the spacecraft's solar panels opened and the onboard equipment became activated. Then, as it flew above Russian territory later in this orbit, controllers checked the equipment aboard *Zvezda* and oriented it toward the Sun so that it could begin to generate a buildup of solar power. On the next flyover, technicians actuated the spacecraft's start tracker and tested the communications system. The following day, the module's two main engines fired twice, in order to raise its altitude to that of the ISS. The day before docking, controllers maneuvered it into position. Guided by the Kurs rendezvous system, *Zvezda* became the passive vessel, and *Zarya–Unity* pursued, aligned itself with, and captured it. The mating of the two parts occurred on July 25, 2000, after a 25-minute sequence in which the fastening of hooks and latches completed the process, adding *Zvezda*'s 43-ft-long (13 m) body and 98-ft (29.8-m) wingspan to the overall structure. Russian controllers monitoring pressure confirmed that an airtight seal had been achieved.

From this point on, *Zvezda* gradually assumed the role of ISS's nerve center, in time relegating *Zarya* to a storage container for propellant and other materials. *Zvezda*'s three pressurized holds—a round Transfer Compartment forward, a cylindrical Work Compartment in the middle, and a spherical Transfer Chamber aft—served as the domicile for astronauts in the early stages of the ISS (containing sleeping quarters and kitchen, exercise, and personal hygiene areas). Lighted in part by a total of thirteen windows, *Zvezda* served the ISS's need for electrical distribution, data processing (supplied by ESA computers), flight controls, life support, communications, and propulsion. The spacecraft also came equipped with four docking ports, the aft port used to mate with the Russian Progress supply vessels, the forward port attached to *Zarya*.

Key facts

MATED WITH ZARYA-UNITY:	**July 25, 2000**
LENGTH:	**43 ft / 13 m**
WINGSPAN:	**98 ft / 29.8 m**
WEIGHT:	**42,000 lb / 19,000 kg**
LAUNCH:	**July 12, 2000**

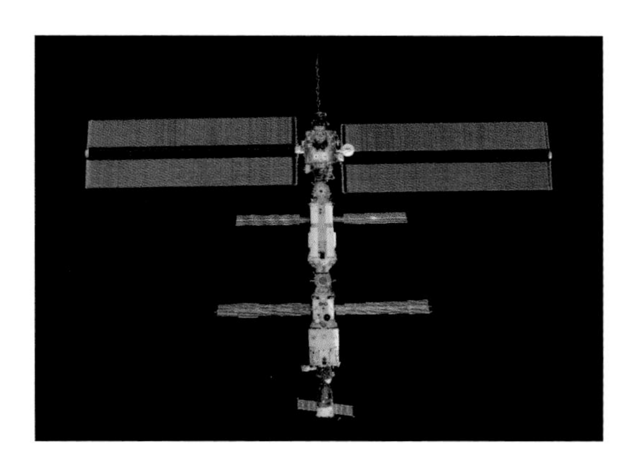

LEFT
The *Unity*, *Zarya*, *Zvezda*, and Soyuz stack (top to bottom). A solar array 240 ft (73 m) long and 38 ft (11.6 m) wide was added to *Zarya* in 2000.

RIGHT
Seen from the Space Shuttle *Endeavour*, the ISS as it looked about a year and a half after construction had begun, including (lower right to upper left) the Soyuz spacecraft, the recently added *Zvezda*, *Zarya*, and the *Unity* module.

PAGES 36–37
The interior of the *Zvezda* module, looking aft.

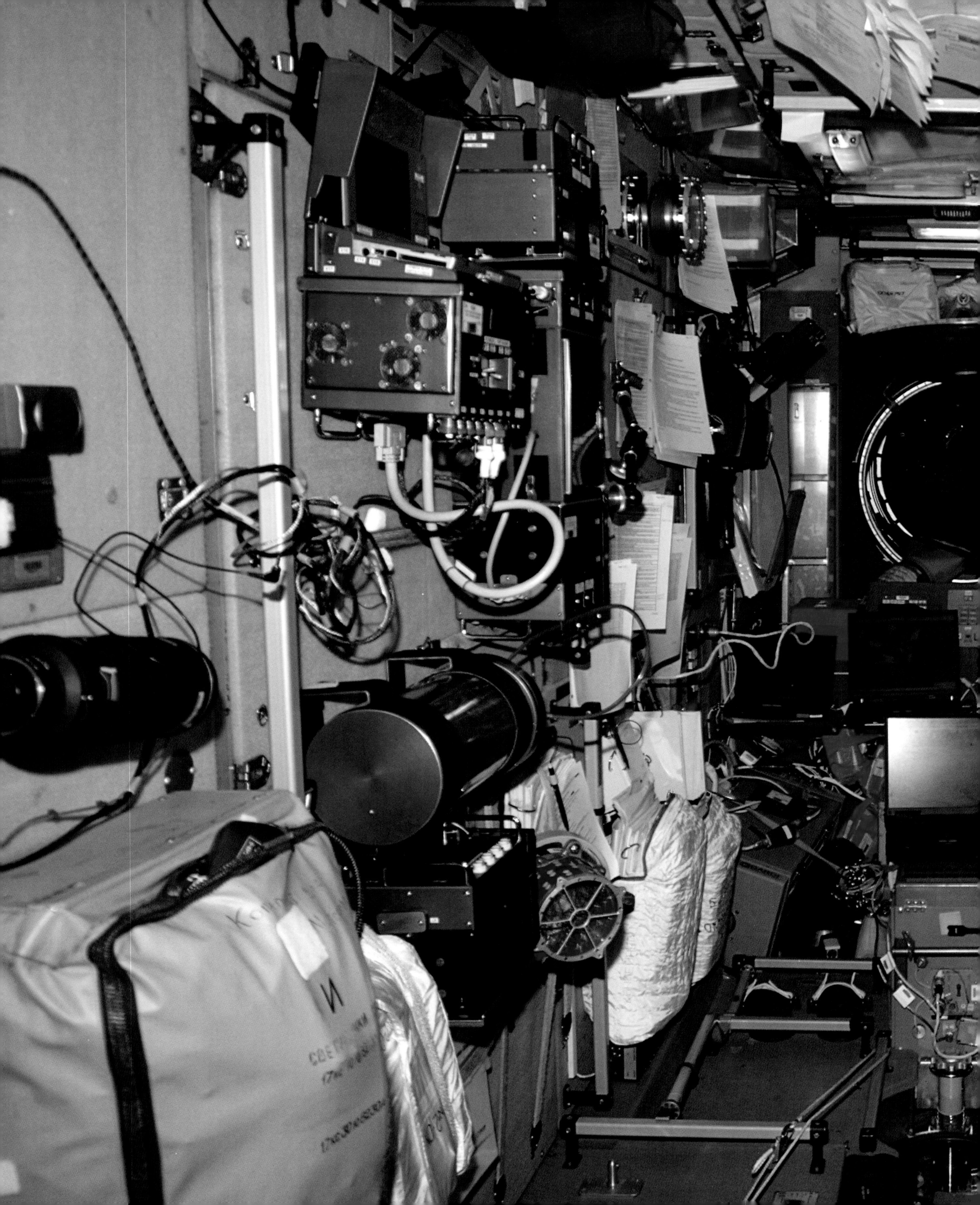

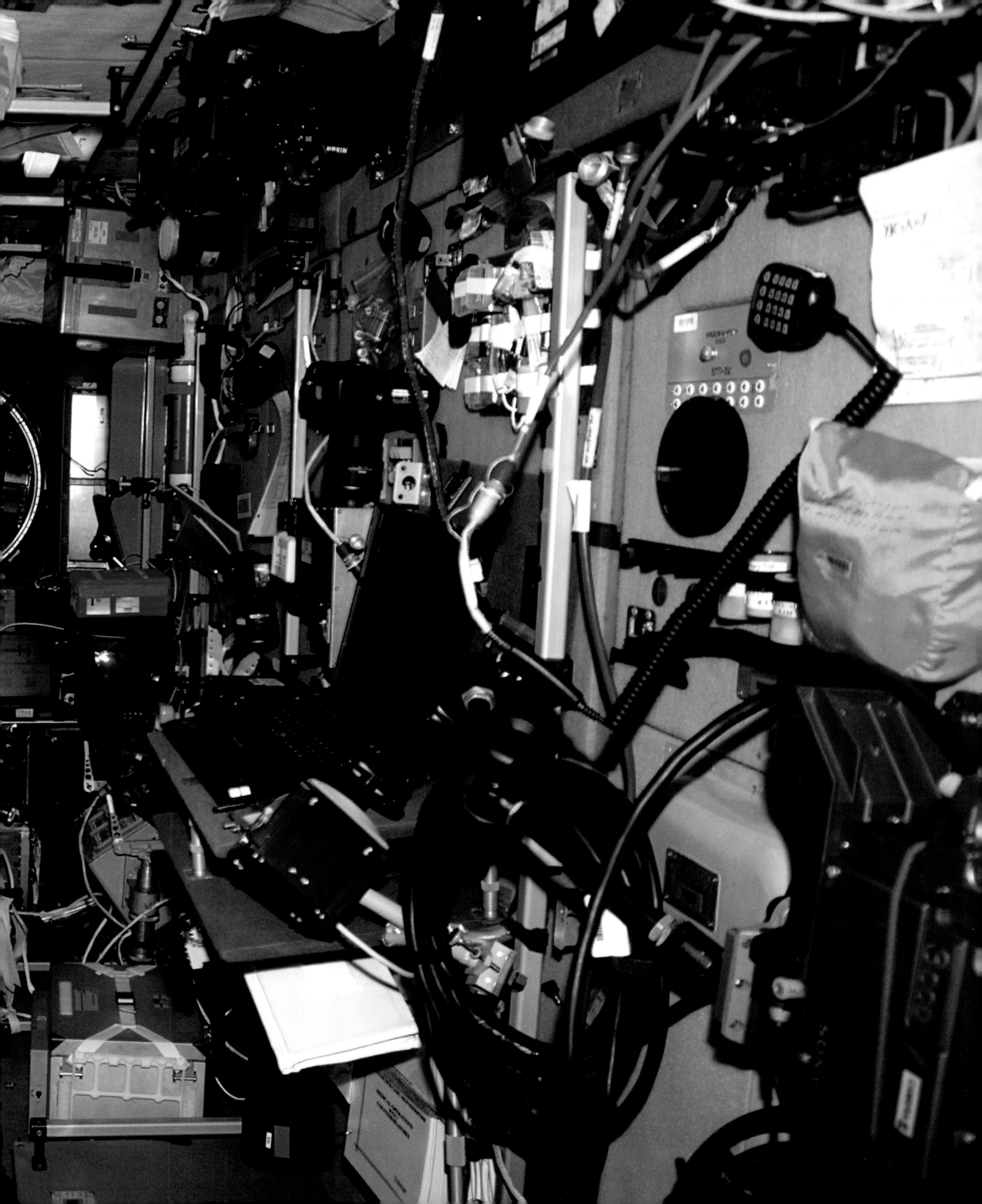

Progress

Less than two weeks after the addition of *Zvezda* to the station, the ISS had its first automated supply visit from an unmanned version of the Soyuz spacecraft. *Progress M-1-3* left Baikonur on August 6, 2000, and reached the ISS on August 8. A few days later, it transferred fuel to the station but remained connected until the crew of mission STS-106 (who arrived at the station on September 8, 2000, on a supply mission) transported *Progress*'s cargo into the ISS. The first Progress voyage to the ISS ended on November 1, 2000, at which time it separated from *Zvezda*, fell out of orbit, and burned up over the Pacific Ocean.

Meanwhile, the ISS's first crew—Expedition 1—left Baikonur on October 31, 2000, aboard the *Soyuz TM-3*. Carrying a three-man complement of Commander Bill Shepherd, Soyuz Commander Yuri Gidzenko, and Flight Engineer Sergei Krikalev, the Soyuz made an automated docking with the *Zvezda* rear port, and the men clambered on to the station on November 2, 2000, for a four-month stay.

But while Soyuz brings the crews, it hardly has the room to satisfy their long-term needs. The permanent habitation of the ISS proved the pivotal importance of the Progress vehicles to the entire station concept. The Russians understood from years of experience that a continuous presence in space demanded a reliable and affordable means of supply. Like much of the Russian contribution to ISS, Progress came from a long and tested progeny. The Soviet space program planned for the contingency of re-supply as long ago as the early 1970s, as an adjunct to the *Salyut-6* station. Designers at TskBEM (predecessor to the spacecraft component manufacturer RKK Energiya) thought the answer lay in an unmanned, non-recoverable spacecraft much like the Soyuz Orbital Module. (For more on the Soyuz spacecraft, see p. 46.) A hold similar to the living quarters of the Soyuz would carry the pressurized cargo. In addition, the unpressurized portion of the Soyuz design—the re-entry module—would serve on the re-supply ship as the container for propellants and refueling. The TskBEM engineering team completed the original design in February 1973, and technicians finished the first production vehicle in November 1977. Given the name *Progress*, it flew initially to *Salyut-6* on January 20, 1978. It succeeded in its mission, as did all of the other forty-two Progresses that supported the *Salyut-6* and *-7* space stations, enabling the Soviets to sustain an almost uninterrupted presence in space.

With a modification of its flight-control systems, in 1986 Progress became Progress-M and began servicing the new *Mir* space station. Fourteen years later, designers at RKK Energiya modified the venerable Progress once more. The engineers placed more propellant tanks in the mid-section, transferring the water tanks that had occupied this space forward to the cargo hold. They ran a "necklace" of twelve external tanks (containing an oxygen-nitrogen atmosphere) around the narrow point between the cargo and propellant compartments. They also added an updated Kurs rendezvous and docking system, as well as a digital flight-control system. In its revised version, *Progress M-1* flew to the aging *Mir* space station in February 2000. Six months later, launched with the familiar Soyuz booster, Progress—in its *M-1-3* incarnation—made its maiden stop at the International Space Station.

As the ISS develops, the familiar outline of Progress in the docking port means life, sustenance, fuel, and cleanliness. Indeed, each spacefarer aboard ISS needs about 66 lb (29 kg) of consumables a day, roughly 2000 lb (907 kg) per passenger per month. The work of supply is carried out by a spacecraft a little more than 57 ft (17.4 m) long and 14 ft (4.3 m) in diameter. But Progress's stature is deceiving. Because it has no passengers, nor such heavy items as heat shields or life-support systems, nor fuel for return landings (except for the de-orbit burn), both space and bulk are conserved. That explains in part why the many variants of Progress have weighed between 15,500 and 16,000 lb (7000–7260 kg) at launch, but the cargo constituted between 5060 and 5500 lb (2300–2500 kg), or an impressive 30 percent of the total mass. Progress-M added an extra cargo feature that enabled objects to be returned to the Earth. Its Raduga capsule, weighing 770 lb (349 kg) itself, can carry a 330-lb (150-kg) payload to a safe landing. Finally, unlike the initial version of the spacecraft, Progress-M generates power from solar arrays measuring about 35 ft (10.6 m) from end to end.

LEFT
The *Progress 22* supply ship—packed with 5000 lb (2270 kg) of oxygen, fuel, food, and water—approaches the ISS in June 2006.

RIGHT
***Progress 22* docked to the ISS, viewed from a window on the station. The Earth looms in the background.**

Key facts

MANUFACTURER:	RKK Energiya
CARGO:	Up to 5500 lb / 2500 kg of supplies
LENGTH:	57 ft / 17.4 m
DIAMETER:	14 ft / 4.3 m
WEIGHT:	Up to 16,000 lb / 7260 kg (with cargo)
FIRST LAUNCH TO ISS:	August 6, 2000

ABOVE
A close-up of *Progress 20* as it undocks from the ISS in June 2006.

LEFT
***Progress 22* undocks and withdraws from the ISS, carrying trash and unneeded equipment, soon to burn up in the Earth's atmosphere. A replacement ship, *Progress 24*, reached the ISS a few days later.**

Destiny Laboratory

ABOVE
The Shuttle *Atlantis* as seen from above by an ISS crewmember in February 2001. The U.S. *Destiny* Laboratory sits in the cargo hold, just prior to docking.

ABOVE, CENTER
The Canadarm grips the *Destiny* module and lifts it out of *Atlantis*'s payload bay.

ABOVE, RIGHT
Two faces in the crowd: astronauts Susan Helms and James Voss look out from *Destiny*, photographed on an EVA during STS-100.

RIGHT
Japanese astronaut Soichi Noguchi navigates across the *Destiny* Laboratory during an EVA outside *Discovery*. The mission, in August 2005, was the first following the loss of *Columbia*.

On February 7, 2001—their 100th day in orbit, with one more month to go—ISS Expedition 1 crewmembers Shepherd, Gidzenko, and Krikalev awoke to the excitement of a new module about to arrive at the station. On that date, Space Shuttle *Atlantis* (mission STS-98) left Kennedy Space Center with the American *Destiny* Laboratory in its cargo bay. Meanwhile, the cosmonauts and astronaut aboard ISS tidied up their living quarters before the Shuttle's arrival, sending out a Progress ship packed with their trash for immolation in the Earth's atmosphere. After two days of pursuing the ISS, the Shuttle docked with it. During *Atlantis*'s week-long stay, its astronauts conducted three spacewalks to mate *Destiny* to the forward part of *Unity* (during which time *Atlantis*'s engines raised the orbit of the ISS by over 82,000 ft/25,000 m). The three ISS crewmen and the four men and one woman aboard the Shuttle greeted one another cheerfully as they all floated into the brilliant white interior of *Destiny*, after which they connected equipment that made the new module the hub of ISS life support.

When *Atlantis* departed, it left behind the first ISS module devoted to the mission of the station, rather than to its infrastructure. It also left behind a space station bigger than any yet in orbit: 171 ft (52 m) long, 90 ft (27 m) high, 240 ft (73 m) wide, and weighing 224,000 lb (101,600 kg). The first of the ISS's science components, the $1.4 billion *Destiny* would later be linked to Russian, Japanese, and European laboratories that also enabled scientific research aloft. Built by the Boeing Company at Marshall Space Flight Center and fashioned from aluminum, the module has a debris-shielding blanket (composed of material similar to that in bullet-proof vests) just below its outer shell. *Destiny* weighs 32,000 lb (14,500 kg), measures 28 ft (8.5 m) in length and 14 ft (4.2 m) in diameter, and at each end has two cones with hatches that allow crews to enter and leave. A deceptively simple-looking cylinder from the outside, it has insides containing more than enough complexity to make up the difference. It is the size of a large business jet, and its rectangular interior consists of four zones, or rack faces, each holding six racks, twenty-four in all for the entire spacecraft. The racks themselves—73 in. (185 cm) tall and 42 in. (106 cm) wide—swing outward for access. Of the twenty-four racks, thirteen offer space for scientific experiments, while the remaining eleven furnish life-sustaining necessities: generating electrical power, cooling and purifying water, controlling temperature and humidity, and renewing the station's atmosphere. *Destiny* also features an optical quality window, through which crewmembers assist Earth-bound meteorologists and geologists in photographing and videoing, in very high quality, fires, floods, avalanches, coral reefs, and other large natural features.

Destiny undertakes a broad range of scientific experiments. Inside the racks, scientists may place electrical and fluid connectors, sensors, video cameras, motion dampeners, and whatever else their research requires. Using this and other equipment, experimenters from around the world, in such fields as biomedicine, physics, biology, chemistry, Earth science, and ecology, study the effects of zero gravity on physical and biological processes. *Destiny* also makes possible research about human physiology in space, in order to assess the practicality of future missions beyond the Earth.

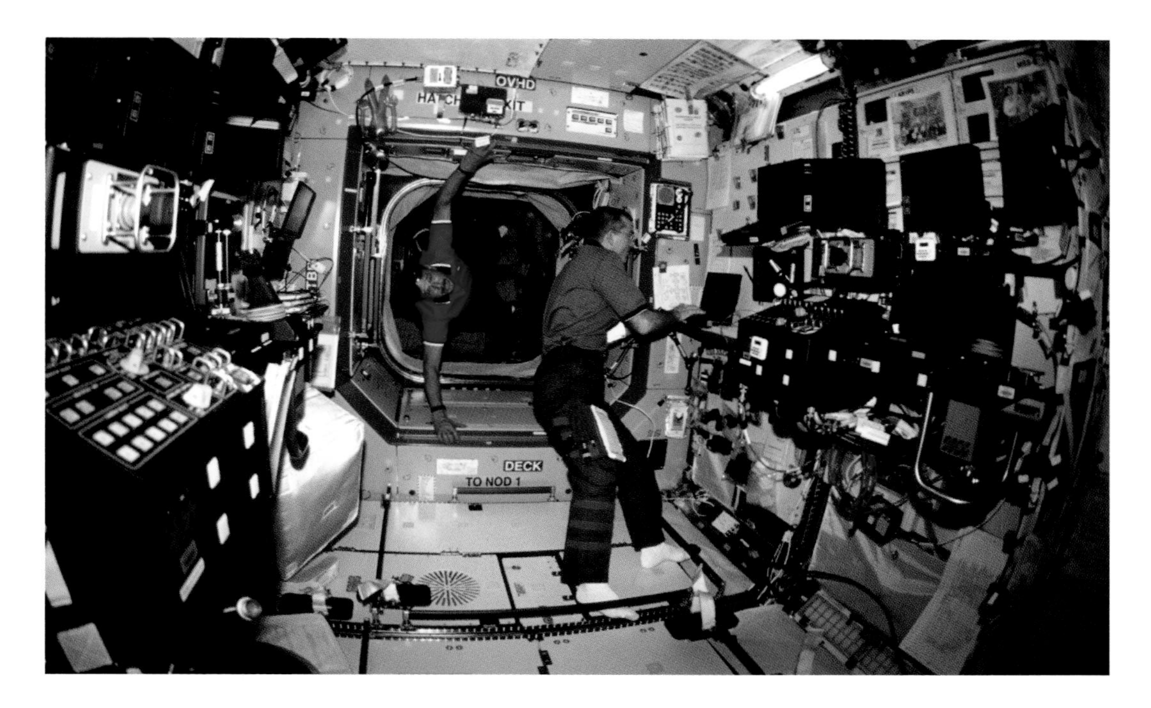

Key facts

MANUFACTURER:	**Boeing Company**
OVERALL STRUCTURE:	**Aluminum**
COST:	**$1.4 billion**
LENGTH:	**28 ft / 8.5 m**
DIAMETER:	**14 ft / 4.2 m**
WEIGHT:	**32,000 lb / 14,500 kg**
LAUNCH:	**February 7, 2001**

LEFT
***Destiny*'s interior as it looked to Expedition 1 and STS-98 crews, early in its habitation.**

ABOVE
A fully functional *Destiny* as it appeared in August 2001, showing astronaut James Voss inside the module, and Commander Scott Horowitz floating through the hatch from *Unity*.

Canadarm 2

Like sentries on watch, the parallel Canadarm 2 and the original Canadarm loom over the Earth.

Astronaut Stephen Robinson on an EVA during STS-114 (the Shuttle's return-to-flight mission), connected to Canadarm 2.

Canadarm 2 (top) and the original Canadarm, in service to the Shuttle *Discovery* and a Soyuz spacecraft, both docked to the ISS in 2005.

BELOW
The newly installed Canadarm 2 on the ISS, 2001.

RIGHT
Swedish astronaut Christer Fuglesang balances on the foot restraint of Canadarm 2 during STS-116 in 2006.

FAR RIGHT
Canadarm 2 against a background of the Caribbean Sea and Hurricane Emily in July 2005.

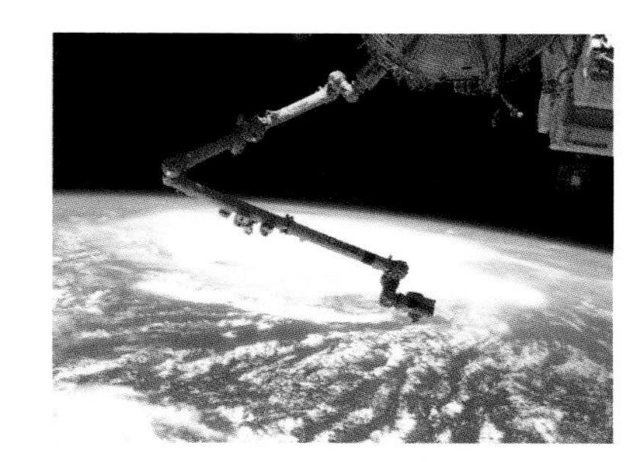

A little more than two months after *Destiny* transformed the function and functioning of the ISS, another key ingredient arrived at the station aboard a Shuttle flight. By this time, the Canadarm had proved itself for twenty years on more than fifty Shuttle missions. This success prompted the ISS partners and the Canadian Space Agency to agree upon the development of a different but related piece of equipment for the ISS, called the Space Station Remote Manipulator System—or, more familiarly, Canadarm 2. During the five years between 2006 and 2011, ISS plans called for transporting heavy pieces to the station, and Canadian scientists and engineers devised a novel machine by which to maneuver and assemble these parts in orbit.

Canadarm 2 arrived at the station aboard the Space Shuttle *Endeavour* after being launched from Kennedy Space Center on April 19, 2001. Using the Shuttle's Canadarm, Pilot Jeff Ashby removed Canadarm 2 from the payload bay and attached it to a special platform on *Destiny*. Canadian astronaut Chris Hadfield, in concert with American astronaut Scott Parazynski, then went on a spacewalk during which they unbolted Canadarm 2 and attached power to it. Two days later, the Hadfield–Parazynski team again walked in space, this time installing a handhold for Canadarm 2 on *Destiny* (the ISS component containing the arm's workstations). Unlike its predecessor, Canadarm 2 did not pivot from a permanent fixed point. Incredibly, it moves by gripping a handhold, vaulting end-over-end like a slow-motion gymnast, and grasping another handhold. Appearing oddly surreal as it cartwheeled across the ISS, Canadarm 2 nonetheless enjoyed great success in serving the construction needs of the ISS crews.

Canadarm 2 surpasses the original Canadarm in a number of respects. It is equipped with seven motorized joints rather than three, weighs 3620 lb (1642 kg) compared to 905 lb (410 kg), has greater maneuverability, incorporates sensors that give its operators a feeling of touch, stretches 8 ft (2.4 m) longer, has more advanced computer controls, and has the capacity to lift heavier and larger objects; it can even assist with docking the Space Shuttle.

The value of Canadarm 2 increased even more when *Atlantis*, in mission STS-111, delivered the arm's Mobile Base System (MBS) to the ISS in June 2002. A small truck-like platform attached to rails that run the length of the ISS, MBS tethered one end of Canadarm 2 as it traveled. Not only did this system enable the arm to range over the full length of the station, but it also enabled it to move more quickly than in its worm-like incarnation, as well as to achieve lateral mobility as it navigated the station's main trusses.

Key facts

LENGTH:	**58 ft / 17.4 m**
WEIGHT:	**3620 lb / 1642 kg**
MOTORIZED JOINTS:	**7**
LAUNCH:	**April 19, 2001**
MOBILE BASE SYSTEM DELIVERED TO ISS:	**June 2002**

Quest and Pirs

LEFT
As the drama of the Earth's weather unfolds below, the *Soyuz TMA-6* spacecraft (identified here by its solar arrays) is docked to Pirs, which in turn is secured to the ISS.

RIGHT, TOP
The Russian-made Pirs Docking Compartment connects the *Progress 20* supply ship to the ISS in June 2006.

RIGHT
ESA's Roberto Vittori (left) and South African space tourist Mark Shuttleworth afloat inside the Pirs airlock.

Key facts

QUEST JOINT AIRLOCK MODULE ARRIVAL AT ISS:	**July 2001**
PIRS DOCKING COMPARTMENT ARRIVAL AT ISS:	**September 2001**
QUEST MANUFACTURER:	**Boeing Company**
PIRS MANUFACTURER:	**RKK Energiya**
QUEST DIMENSIONS:	**18 ft / 5.5 m long, 13 ft / 4 m in diameter**
PIRS DIMENSIONS:	**16 ft / 4.9 m long, 8.4 ft / 2.6 m in diameter**

During the summer of 2001, the ISS again increased in size and bulk, with the addition of two critical adapters needed to broaden access to space and to accommodate the different spacesuits worn by cosmonauts and astronauts—a part of spaceflight in which commonality did not yet exist. First came the U.S. Quest Joint Airlock Module in July 2001. Until its installation, an awkward situation existed in which astronauts could access space only through airlocks on the Space Shuttle, and cosmonauts could only ingress and egress through *Zvezda*. Quest represented a universal port, enabling spacewalks with either Russian or American suits. Fabricated at the Boeing Company's facilities at the Marshall Space Flight Center in Huntsville, Alabama, this aluminum module—18 ft (5.5 m) long, 13 ft (4 m) in diameter, and weighing 13,368 lb (6064 kg)—became the primary airlock for the ISS. It operated like the Shuttle airlock, but with major modifications (the newer unit enabled spacewalks without venting any air). Its two cylindrical chambers, attached by a bulkhead, performed two functions: one served as a warehouse for spacesuits, adapters, and related equipment; the other acted as a chamber in which pressure could be increased or decreased prior to an EVA. On July 15, 2001, during Shuttle mission STS-104, the crewmembers attached Quest to the starboard side of *Unity*.

Two months after Quest's hook-up, Russia's multi-purpose Pirs Docking Compartment joined the station on September 16, 2001, after launch from a Soyuz rocket. Built by RKK Energiya, it had smaller proportions than Quest—16 ft (4.9 m) long, 8.4 ft (2.6 m) in diameter, and weighing 8000 lb (3600 kg) —but a wider variety of functions. If Quest's airlock drew inspiration from that of the Shuttle, Pirs's design paraphrased the docking mechanism of the *Mir* space station. Pirs had three main functions: as a docking port for Progress-M and Soyuz TMA spacecraft, located on the bottom side of *Zvezda*; as an airlock, accessible only to Russian spacesuits; and as a fuel-transfer system, transporting propellant from visiting Progress spacecraft to *Zarya* or *Zvezda*, or from *Zarya* and *Zvezda* to Soyuz or Progress vehicles.

The Pirs Docking Compartment undergoes ground preparations prior to its launch in September 2001.

Astronaut Piers Sellers leaves the American-made Quest Airlock for a spacewalk during STS-112.

Soyuz spacecraft

After the *Columbia* accident in February 2003, all Shuttle missions to the ISS ended, and did not resume until July 26, 2005. (See the section on the Space Shuttle, pp. 16–21.) Because of the Shuttle's carrying capacity, this meant that the station could not increase in size significantly until the American spacecraft resumed operations. But the ISS did continue to function and its crews continued to inhabit it, thanks to the time-honored Soyuz spacecraft and the Russian Federal Space Agency (RKA). The individuals aboard the ISS at the time of the disaster had supplies to last until June 2003. Cosmonauts Yuri Malenchenko and Alexander Kaleri, along with astronaut Ed Lu, had been chosen as the crew of the seventh and succeeding expedition to the ISS, scheduled for launch on the Shuttle *Atlantis* on March 1. With the grounding of the Shuttle, Russian space officials announced with NASA a 185-day mission for Malenchenko and Lu only, to conserve resources on the station. Launched on April 26, 2003, the Soyuz TMA spacecraft's two-day autonomous flight succeeded in bringing Malenchenko and Lu to the station, after which Soyuz evacuated crew number six back to Baikonur. In October, Soyuz again filled in, as the Shuttle underwent further investigation and correction, this time bringing home crew number seven. After a week of transfer operations, crew number eight assumed command. This cycle continued on the same six-month timetable without Shuttle assistance not until July 2005, when *Discovery* returned the Shuttles back to service, but until July 2006, at which time *Discovery* brought ESA astronaut Thomas Reiter to the ISS to join crew number thirteen and stay on for number fourteen. In between, from crew number seven to crew number thirteen, the RKA and Soyuz made fourteen successful flights that kept the ISS occupied and operating during the entire crisis.

The Soyuz TMA spacecraft consisted of three parts connected end to end: a Descent Module in the middle, an Instrumentation/Propulsion Module on one end, and an Orbital Module on the other. In all, it measured about 23 ft (7 m) in length and 9 ft (2.7 m) in diameter; had a span of 35 ft (10.7 m); and weighed 15,884 lb (7205 kg). After insertion into orbit, the Orbital Module affords the crew increased living space, as well as key systems related to rendezvous and docking with the station. The Instrumentation/Propulsion Module contains the primary guidance and navigation controls for the Soyuz. On the way back to Earth, the Instrumentation/Propulsion Module's engines fire a de-orbit burn, at which time both the Orbital and the Instrumentation Modules fall away, leaving the Descent Module to transport the crew home. Using a secondary guidance and navigation system, the pilot steers the module into the atmosphere, and, fifteen minutes before landing, opens four parachutes. Then, just a second before touchdown, the pilot fires the engines to cushion the impact as the Soyuz touches down on the broad steppes of Kazakhstan.

STS-105 Commander Scott Horowitz (center) and mission specialist Daniel Berry pose within the tightly packed interior of a Soyuz spacecraft.

With Earth's horizon in the background, a Soyuz spacecraft is held firmly to the ISS by the Pirs Docking Compartment.

A Soyuz spacecraft, looming large in the window of the ISS and guided by Commander Mikhail Tyurin, prepares to dock with *Zvezda* in September 2006.

A distant Soyuz spacecraft aligns itself with the ISS prior to docking. On this particular mission, the crewmembers were accompanied by space tourist Anousheh Ansari.

Key facts

LENGTH:	23 ft / 7 m
DIAMETER:	9 ft / 2.7 m
SPAN:	35 ft / 10.7 m
WEIGHT:	15,884 lb / 7205 kg

Multi-Purpose Logistics Module

The MPLM *Leonardo* holds fast to the cargo bay of the Space Shuttle *Discovery*.

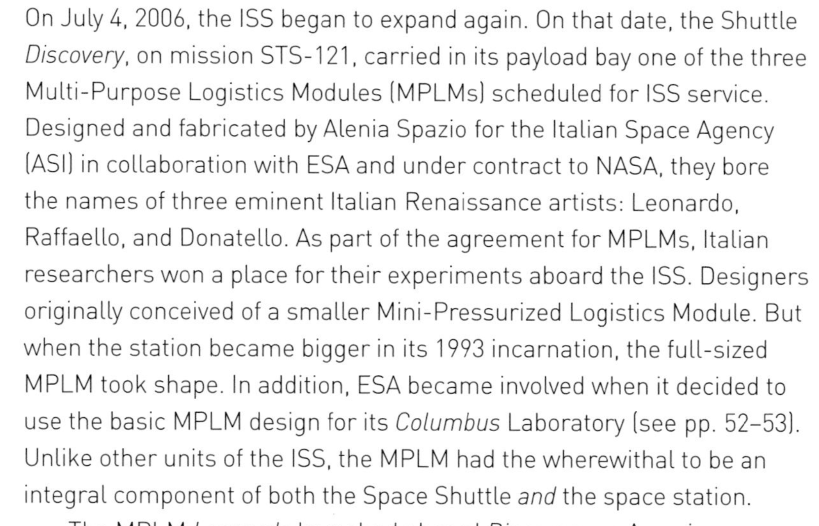

Key facts

MANUFACTURER:	**Alenia Spazio for the Italian Space Agency**
LENGTH:	**21 ft / 6.4 m**
DIAMETER:	**15 ft / 4.6 m**
WEIGHT:	**9000 lb / 4100 kg**
LAUNCH:	**July 4, 2006**

On July 4, 2006, the ISS began to expand again. On that date, the Shuttle *Discovery*, on mission STS-121, carried in its payload bay one of the three Multi-Purpose Logistics Modules (MPLMs) scheduled for ISS service. Designed and fabricated by Alenia Spazio for the Italian Space Agency (ASI) in collaboration with ESA and under contract to NASA, they bore the names of three eminent Italian Renaissance artists: Leonardo, Raffaello, and Donatello. As part of the agreement for MPLMs, Italian researchers won a place for their experiments aboard the ISS. Designers originally conceived of a smaller Mini-Pressurized Logistics Module. But when the station became bigger in its 1993 incarnation, the full-sized MPLM took shape. In addition, ESA became involved when it decided to use the basic MPLM design for its *Columbus* Laboratory (see pp. 52–53). Unlike other units of the ISS, the MPLM had the wherewithal to be an integral component of both the Space Shuttle *and* the space station.

The MPLM *Leonardo* launched aboard *Discovery* on American Independence Day in 2006 gave the appearance of a large pressurized cylinder, approximately 21 ft (6.4 m) long and 15 ft (4.6 m) in diameter. It weighed about 9000 lb (4100 kg) and could carry roughly 20,000 lb (9100 kg) of cargo on a system of sixteen standard racks. The Minus Eighty Degree Laboratory Freezer for ISS (MELFI), built by ESA, exemplified the type of items MPLMs handled. MELFI provided cold storage for various scientific samples until they could be transported back to Earth. Once docked to the ISS's *Unity* module by the Canadarm, the crew disgorged *Leonardo* of supplies, equipment, and experiments. In return, *Leonardo* carried waste and completed experiments back to Earth when it returned to the Shuttle cargo hold prior to the flight home. But MPLMs needed to be more sophisticated than mere celestial moving vans. Because they also held human cargo during their periods as space-station components, they required life support, fire suppression, electrical distribution, and computing power, all to create livable space for two human beings.

ABOVE
An overhead shot of MPLM *Leonardo*, safely anchored to the Shuttle *Discovery*'s cargo bay.

BELOW, LEFT
Once *Leonardo*'s mission had been accomplished during STS-121, astronauts Stephanie Wilson and Lisa Nowak maneuvered the MPLM away from its docking port on *Unity*.

BELOW, RIGHT
Astronauts Wilson and Nowak, located inside the *Destiny* module, transfer *Leonardo*, now detached from *Unity*, back to *Discovery*'s payload compartment.

Automated Transport Vehicle

The year after the return of the Shuttles to full capacity in 2006–07 was pivotal for the evolution of the ISS. Until this point, its assembly had been halting; after it, the station assumed a size and structure closer to its final configuration. The launch program of 2008 included a major European contribution called the Automated Transfer Vehicle (ATV), the first of which is called *Jules Verne* in honor of the famous French novelist. Conceived and fabricated by the European Space Agency, the ATV may constitute ESA's most advanced spacecraft to date. It shares the task of supplying the ISS with the Russian Progress spacecraft and the Italian MPLMs. Unlike its counterparts, the ATV (although unoccupied on its flights to the station) links by Russian-made docking devices to the Russian Service Module and remains as a component of the ISS for up to six months. There, in a pressurized atmosphere, astronauts and cosmonauts obtain supplies (spare parts, experimental materials, water, air, and food) and store refuse. Although based on the MPLM's cylindrical design, the ATV, in contrast, behaves as a true spacecraft, with the capacity for automated approach and docking, and its own navigation and propulsion systems.

The ATV has the capacity to haul 16,600 lb (7500 kg) to the ISS, and 14,400 lb (6500 kg) on the return flight (which ends in immolation in the atmosphere). Weighing 40,600 lb (18,400 kg), it measures a little less than 34 ft (10.3 m) in length, almost 15 ft (4.6 m) in diameter, and receives power from a solar array about 73 ft (22.3 m) across. Inside, it carries eight standard racks for cargo and experiments, as well as tanks of air, water, and propellant. The ATV flies from ESA's spaceport in Kourou, French Guiana, atop a dedicated Ariane 5 rocket, and requires three days to make contact with the ISS. ESA planned to construct a minimum of seven ATVs, the product of a consortium led by the European Aeronautic Defence and Space Company (EADS) and consisting of thirty firms from ten European countries. Eight other American and Russian companies also contributed. Testing and integration of the many pieces and systems took place in The Netherlands, while the Ariane 5 construction and the ATV launch control activities were undertaken in Toulouse. Patrice Amadieu, ATV Deputy Program Manager, noted the technical and multinational complexity that was involved with the spacecraft:

> ATV is a complex programme; we have to work on a lot of interface aspects with the Russians, because we'll dock with their module; with NASA who [*sic*] has the overall responsibility for the ISS; with Ariane 5 engineers and with the ATV control centre in Toulouse. So it is much more convenient to be at the same location as our main contractor [EADS]. We can easily work out issues with our industrial partners in preparation of a meeting with the Russians or with the Americans.[4]

The first ATV met with success when it lifted off atop a modified Ariane 5 rocket on March 9, 2008. It then accomplished a series of complex maneuvers prior to docking with the ISS.

FAR LEFT
RSC Energia engineer Vladimir Syromiatnikov, designer of the ATV's docking system, stands beside his invention.

LEFT
The ATV's pressure module, in which ISS crews work once the supply ship has docked with the station, on the factory floor at EADS Space Transportation in Bremen, Germany.

Key facts

PRIME CONTRACTOR:	**EADS (for ESA)**
CARGO CAPACITY:	**16,600 lb / 7500 kg going to the ISS, 14,400 lb / 6500 kg leaving it**
LENGTH:	**34 ft / 10.3 m**
DIAMETER:	**15 ft / 4.6 m**
SOLAR ARRAY:	**73 ft / 22.3 m**
WEIGHT:	**40,600 lb / 18,400 kg**

BELOW
Once the craft is in orbit, the protective fairing breaks away; the ATV is then awakened by ground controllers.

RIGHT
An artist's rendering of ESA's ATV inside its fairing while being boosted into orbit aboard an Ariane 5 rocket.

BELOW, RIGHT
A drawing of the ATV in its activated mode. Once docked to the ISS, it forms an integral part of the station for as long as six months.

Columbus Laboratory

An artist's conception of the *Columbus* Laboratory, once installed on the ISS.

A cutaway drawing of the *Columbus* Laboratory, aboard the ISS.

Key facts

PRIME CONTRACTOR:	**EADS (for ESA)**
CREW:	**Three**
PAYLOAD RACKS:	**15**
LENGTH:	**23 ft / 7 m**
DIAMETER:	**15 ft / 4.6 m**
ON-ORBIT WEIGHT (WITH FULL PAYLOAD):	**42,000 lb / 19,000 kg**

Hoisted by crane at the Kennedy Space Center's Space Station Processing Facility, *Columbus* is lowered on to a work stand.

***Columbus* inside a container prior to being transported by an Airbus Beluga aircraft to Kennedy Space Center in Florida for pre-launch processing.**

If the ATV represents one of the most complex spacecraft yet attempted by ESA, the *Columbus* Laboratory—named for the famous Genoese mariner—represents the consortium's most noteworthy long-term contribution to the ISS itself. Yet another spacecraft indebted to the MPLM for its basic design and life-support equipment, *Columbus* differed in that once attached after delivery from the Shuttle, it became a permanent part of the ISS, with a life expectancy of at least ten years. After completion in May 2006 by EADS (the prime contractor) and a ceremony in Bremen, Germany, at which it was presented to ESA, *Columbus* left for Kennedy Space Center in Florida in preparation for its launch on February 7, 2008, aboard the Shuttle *Atlantis*.

The new module has the novel characteristic of enabling researchers on the ground to conduct and oversee experiments in weightlessness from special monitoring centers, and even from their home computers. Holding ten standard payload racks at first (and later fifteen, with the addition of five designated for NASA), each of *Columbus*'s racks—roughly a phone booth in size—provides independent power and cooling for its contents. With some occasional assistance from astronauts or cosmonauts aboard the ISS, Earth-bound researchers in the life sciences, fluid physics, materials, and other fields will be able to conduct thousands of tests from this new outpost of scientific discovery. For instance, a Fluid Science Laboratory will study liquids in weightless conditions, with possible applications for optical lens manufacture and the cleanup of oil spills. Moreover, four platforms on the outside of *Columbus* will enable mission controllers in Germany to observe the effects of direct exposure to the vacuum of space on a host of experiments, such as the capacity of bacteria to survive on meteorites.

With an outer shell fashioned from aluminum, *Columbus* underwent fabrication in Bremen, Germany, by Astrium GmbH, which led many subcontractors. The module supports a crew of three in a vehicle about 23 ft (7 m) long and 15 ft (4.6 m) in diameter. Without payload, it weighed more than 22,000 lb (10,000 kg). Fully loaded for launch, its bulk increased to 28,000 lb (12,700 kg). With maximum weight on orbit, its bulk grew to 42,000 lb (19,000 kg), nearly 20,000 lb (9100 kg) of which constituted payload. Among the experiments packed tightly into its cargo hold, *Columbus* hosted the Biolab rack, which served as a center for research related to micro-organisms, plants, cells, small invertebrates, and tissue cultures. Its sponsors hoped to answer the provocative and fundamental question that will decide whether human beings are physiologically capable of long-duration space travel: what effect does microgravity have on life, from cells to such complex creatures as men and women?

In exchange for the launch of *Columbus* by NASA, ESA representatives agreed to manufacture the second "juncture box" of the ISS, known as Node 2. The Italian Space Agency in Turin took charge of the project as prime contractor. Upon its completion, ESA officials transferred ownership to NASA in June 2003. Eventually, each of the three international laboratories—*Destiny*, *Columbus*, the Japanese Experiment Module (see pp. 54–55)—will be docked to Node 2's stubby frame (23 ft 6 in./7.2 m long and 14 ft 6 in./4.4 m in diameter), which will bind them all together.

The Japanese Experiment Module

The Japanese Experiment Module (JEM), also known as *Kibo* (meaning "hope"), is the contribution of the Japanese Aerospace Exploration Agency (JAXA) to the ISS and represents another big building block for the station. Delayed for several years owing to the *Columbia* accident and funding uncertainties, *Kibo* consists of six parts, requiring three separate Shuttle launches: the Pressurized Module, the Exposed Facility, the Inter-Orbit Communications System, an Experiment Logistics Module with an exposed section, an Experiment Logistics Module with a pressurized section, and a Remote Manipulator System. In addition, the Japanese added a scientific airlock, the gateway from the Pressurized Module to the open space environment. Almost a small station in itself, *Kibo* has enabled JAXA to enter ISS participation in a grand fashion, with an ambitious and complex set of interrelated machines.

The Pressurized Module component remained basically unchanged from the time of its conception in 1985 to its arrival at Kennedy Space Center in May 2003, a rarity in light of the ISS's relentless evolution. The module has much in common with *Columbus* and *Destiny*, ESA and NASA's respective space station laboratories. Cylindrical in shape and packed with research equipment, the Pressurized Module on the JEM is equipped with twenty-three racks—ten for experiments, the rest for power generation, communications, air conditioning, and experiment support. Most of its research involves microgravity experiments, but its experimenters will also study ozone depletion and global warming, and will photograph the galaxies through the world's largest wide-angle X-ray camera.

Nearly 37 ft (11.2 m) long and almost 14 ft 6 in. (4.4 m) in diameter, the Pressurized Module approximates a sightseeing bus in size. Mounted on the lower edge of the Pressurized Module (on the end with the airlock), the Exposed Facility offers an unpressurized pallet in the open space environment. Holding up to ten racks at once, the Exposed Facility measures more than 13 ft (4 m) long, nearly 16 ft 6 in. (5 m) high, and almost 18 ft 6 in. (5.6 m) wide. Rather than venture into space themselves, astronauts are able to add or replace experiments on the Exposed Facility with the Remote Manipulator System, a six-jointed

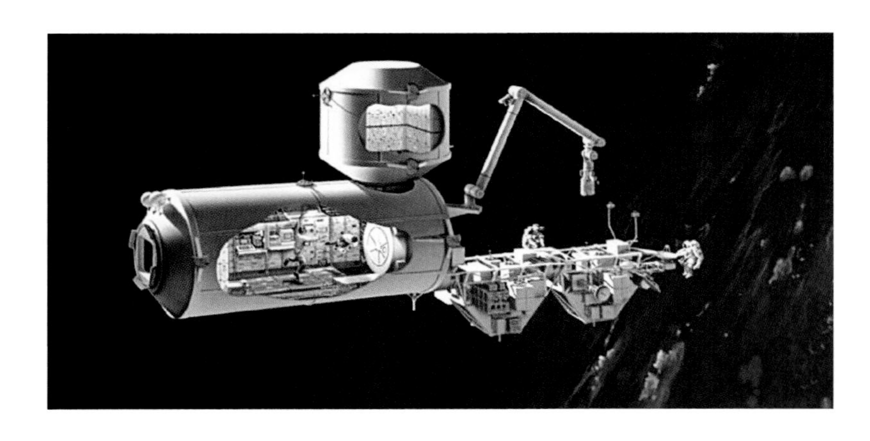

A rendering of the *Kibo* Japanese Experiment Module floating unattached in space.

An artist's conception of the completed *Kibo* Japanese Experiment Module, connected to the ISS.

Key facts

PRESSURIZED MODULE:	**37 ft / 11.2 m long, 14 ft 6 in. / 4.4 m in diameter**
EXPOSED FACILITY:	**13 ft / 4 m long, 16 ft 6 in. / 5 m high, 18 ft 6 in. / 5.6 m wide**
REMOTE MANIPULATOR SYSTEMS (TWO):	**One arm 32 ft 6 in. / 9.9 m long; another 6 ft 2 in. / 2 m long**
EXPERIMENT LOGISTICS MODULES (TWO):	**One 13 ft / 4 m long, 14 ft 6 in. / 4.4 m in diameter; another 16 ft / 4.9 m wide, 7 ft / 2.1 m high, and 14 ft / 4.2 m long**
CONTROL CENTER:	**Tsukuba Space Center**

mechanism consisting of a main arm (32 ft 6 in./9.9 m long, equipped with a television camera) for objects weighing up to 14,000 lb (6350 kg), and a smaller arm (6 ft 2 in./2 m long), attached at the end of the main arm, for lighter tasks.

Among the rest of *Kibo*'s components, the two Experiment Logistics modules predominate. Both store tools, experiment-related equipment, and supplies, but they differ greatly. The pressurized Experiment Logistics Module, shaped like a pill box and attached to the top of the Pressurized Module, holds eight experiment racks within its walls, in a shirt-sleeve environment. It measures nearly 14 ft 6 in. (4.4 m) in diameter and almost 13 ft (4 m) long. The exposed Experiment Logistics Module (attached to the Exposed Facility) contains a pallet capable of holding three experiment racks fully open to the vacuum of space, and measures more than 16 ft (4.9 m) wide, more than 7 ft (2.1 m) high, and almost 14 ft (4.2 m) long. Finally, the Inter-Orbit Communications System enables contact between *Kibo* and ground-based controllers located north of Tokyo at the Tsukuba Space Center, who in turn interface with controllers at the Johnson Mission Control Center in Houston, Texas; the Payload Operations Center in Huntsville, Alabama; and the Russian Mission Control Center, near Moscow.

The ambitious Shuttle launch schedule for 2008 anticipated the delivery of several key components of *Kibo* to the ISS, realized in part in March 2008, when the astronauts aboard *Endeavour* brought the *Kibo* Logistics Module to the station.

Many hands push a trailer holding the Pressurized Module compartment of the *Kibo* Japanese Experiment Module.

Dextre

A drawing of Dextre, as viewed from above, showing its human-like form.

OPPOSITE
Dextre in the role of "Canada Hand," attached to the end of Canadarm 2, which is fixed to the Mobile Base System, which in turn is anchored to the ISS.

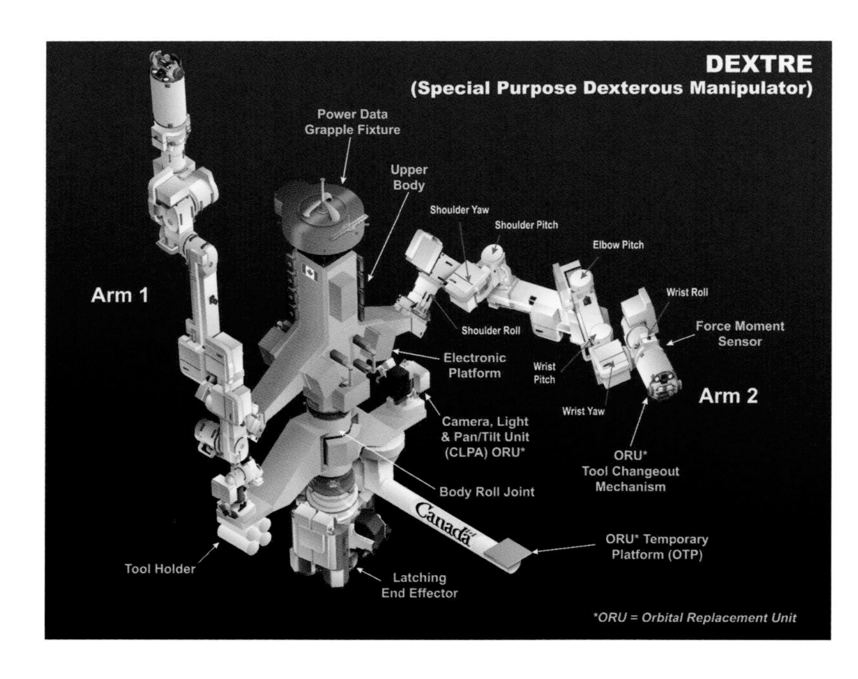

Key facts

PRIME CONTRACTOR:	**MD Robotics**
ARMS (TWO):	**11 ft 6 in. / 3.5 m long, each with seven joints**
LIFT PER ARM:	**1327 lb / 602 kg**
HANDS:	**Computer-guided to sense opposing forces and move smoothly**
HANDS:	**Each equipped with a retractable jaw, illumination, a motorized socket wrench, and a camera**

In yet another step in the development of remote robotic manipulation in space, the Canadian Space Agency followed Canadarm, Canadarm 2, and the Mobile Base System with Dextre, shorthand for the Special Purpose Dexterous Manipulator. Attached more often than not to one end of Canadarm 2 (or alternatively to one of the grapple fixtures on the ISS's sides or to the Mobile Base System), this so-called "Canada Hand" will enable astronauts and cosmonauts to move, connect, and detach small objects requiring light, smooth movements, all from the safety of one of the computer workstations inside the ISS.

Dextre's main contractor—MD Robotics of Brampton, Ontario—led a team of Canadian firms that used Canadarm 2's essential technological concept and applied it to a very different purpose. Enlisting the same design concept as Canadarm and Canadarm 2, the engineers who conceived of Dextre once again used a human anatomical analogy—a tried and well-understood engineering concept—as the blueprint for their design. Viewed from above and to one side, Dextre is shaped like a human being, with a vertical trunk, shoulders, and a pivot at the waist. Extending from the shoulders are two identical arms, each with seven joints and "hands" (known as Orbital Replacement Unit/Tool Changeout Mechanisms, or OTCMs). The OCTMs are highly flexible and operate one at a time to avoid collisions. Computer-controlled and guided, each hand senses the forces opposing its movements and compensates for them automatically, resulting in smooth motion. In fact, designers built this formidable machine to accomplish tasks as refined as inserting a video cassette into a recorder.

The astronaut and cosmonaut corps underwent intensive training on Dextre at the John H. Chapman Space Centre in Saint-Hubert, Quebec. Although the "feel" of Dextre in operation intentionally paralleled that of Canadarm 2, those at the simulated controls of this equipment had much to learn. They discovered how to operate the lights and video equipment, and how to place tools on a dedicated platform and in four tool holders. They learned that each arm (11 ft 6 in./3.5 m long) could lift up to 1327 lb (602 kg). They saw how the parallel retractable jaws on each of the hands gripped objects, as well as the workings of the retractable motorized socket wrench, the illumination system, and the camera mounted on both of the hands. In the end, the aim of these sessions was to enable the ISS's inhabitants to perform such remarkable tasks as installing and extracting batteries, power sources, and computers; connecting electricity and data to payloads; and examining, transferring, and manipulating scientific experiments. Seemingly straightforward, the remote removal of a battery requires complex and precise motions of unbolting and bolting, placement in a tight space, and the hooking up of all the necessary connectors.

Dextre represents the final commitment of the Canadian Space Agency to the ISS, fulfilled in March 2008 when it flew aboard STS-123 and underwent installation by the seven-man crew of *Endeavour*.

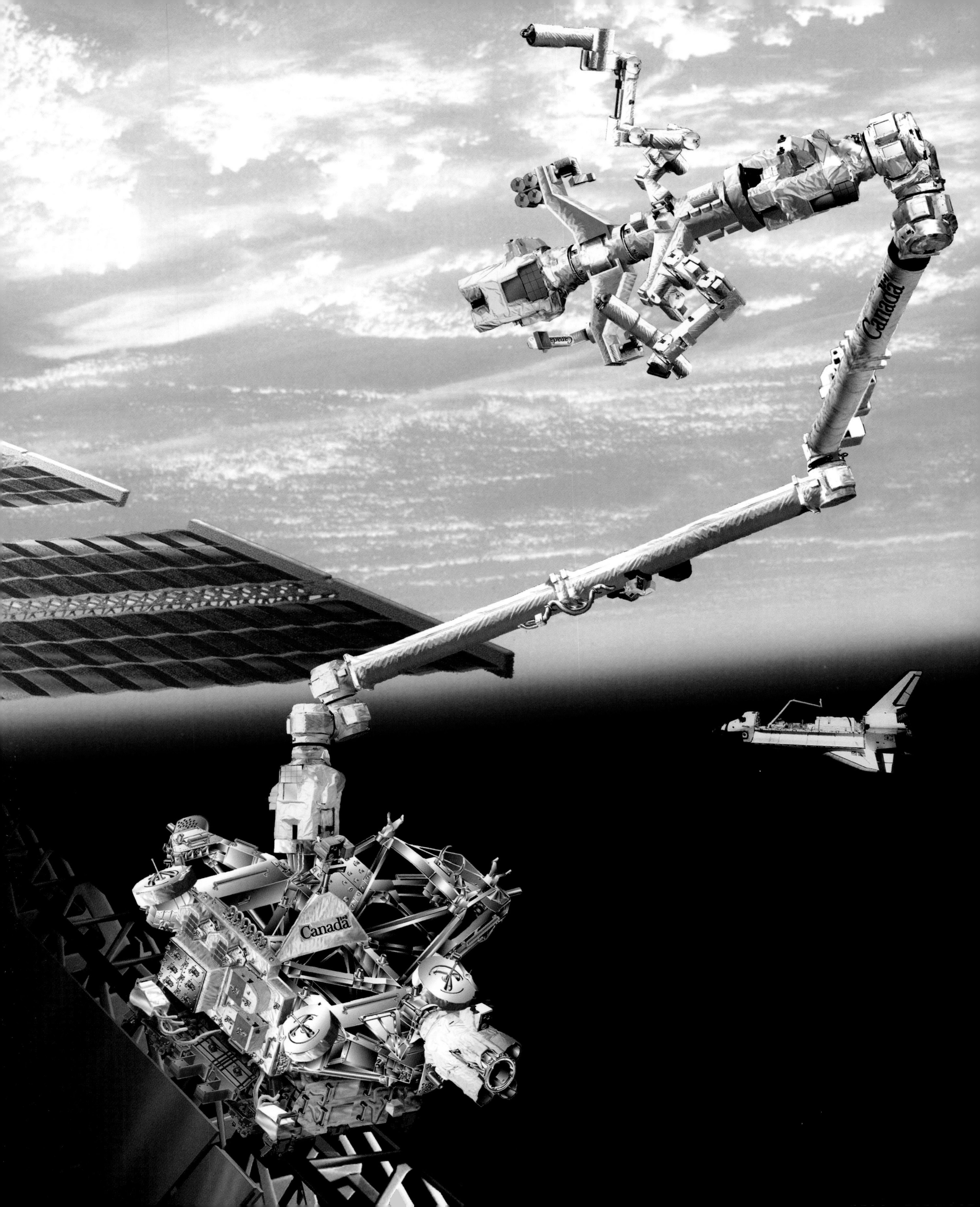
Canada
Canada

Earth Observation

Seeing with New Eyes

2

Introduction

Other aspects of spaceflight may be more appealing because they offer greater drama, but Earth observation reigns supreme in terms of human necessity. Even as NASA struggled from its formative days to mount an offensive against the Soviets in space, it launched the nation's first weather satellite, the polar-orbiting *Tiros 1*, on April 1, 1960. Tiros (Television Infrared Observation Satellite) proved to be a revelation not just to Americans, but also to the world at large. For the first time in history, people were able to observe from above the formation of meteorological patterns, including the disposition of the atmosphere itself, the mechanics of clouds, the motion of storms, and the potential for weather damage by hurricanes, ice, lightning, drought, and all of the other dynamics of the atmosphere. Tiros gave the planet its first chance to observe, if not to predict, the most powerful forces of nature.

A digital photograph of Hurricane Gordon, taken by an International Space Station crewmember on September 15, 2006.

NASA and the National Oceanic and Atmospheric Administration (NOAA) collaborated to conceive and launch a successor to Tiros known as Nimbus, seven of which went into orbit between 1964 and 1978. Nimbus satellites not only took pictures of the clouds, but, equipped with infrared radiometers, also used cameras at night for the first time, enabling scientists to watch the Earth twenty-four hours a day. As Nimbus gathered meteorological data, Landsat satellites monitored the Earth's resources. Since the launch of *Landsat 1* in July 1972—and continuing with more advanced models into the 1980s and beyond—human beings have been in the astounding position of being able to observe every major natural and man-made occurrence on the surface of the land and seas. Such technology facilitated observation of the flow of the oceans, the actions of polar ice, the spewing of volcanoes, the growth of crops, and even the movement of pollution unleashed by humankind.

ABOVE
Earth observation can show routine landscapes and seascapes, but it can also reveal evidence of human tragedy. Here, photographed from the ISS, a plume of smoke rises from New York City on September 11, 2001.

The long, narrow plateau in these images is identified as B-15A, the biggest in the world. A large portion of it broke away from western Antarctica's Ross Ice Shelf in 2000 (see the jagged line bisecting its lower half), an event visible from space.

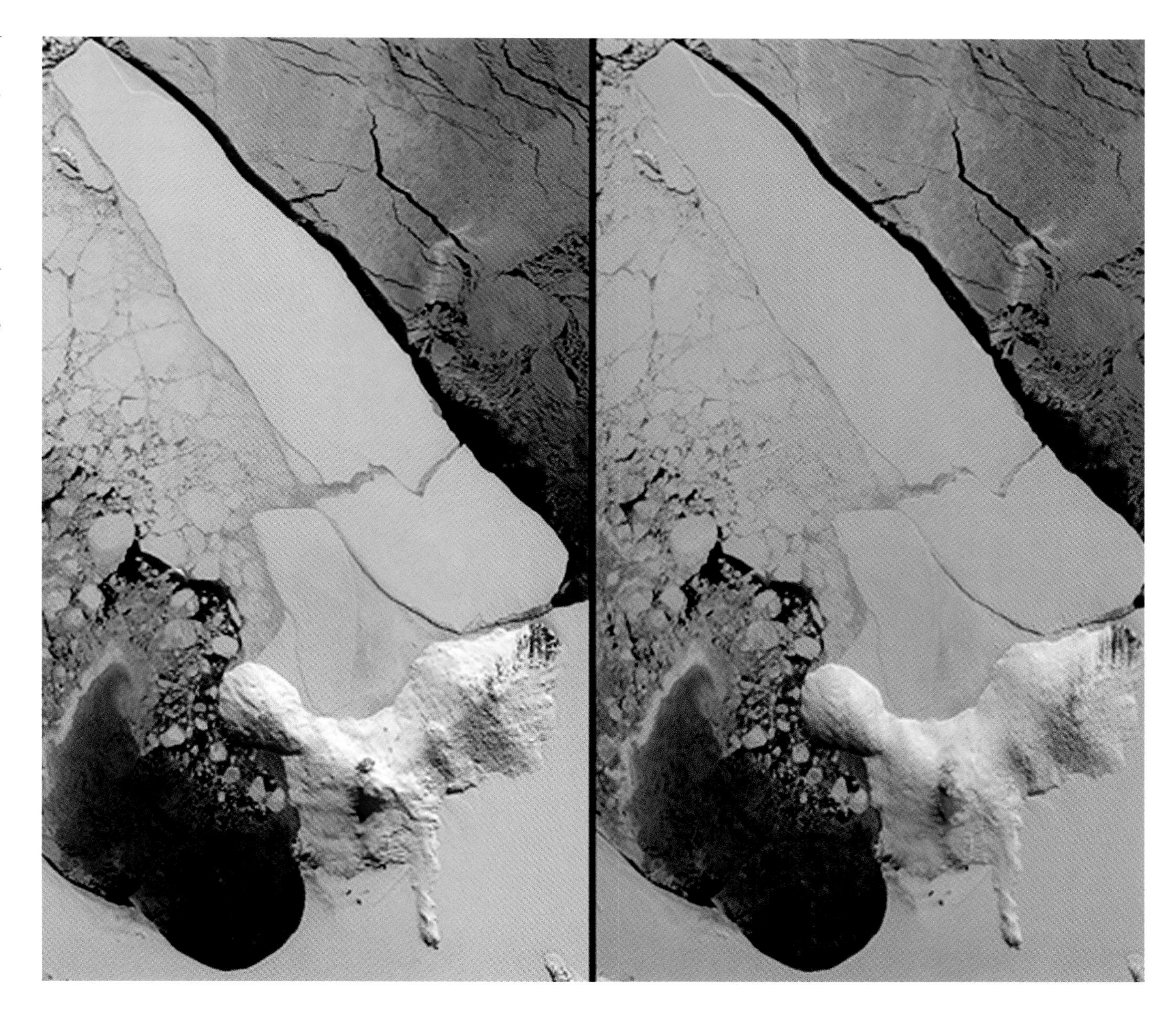

The Earth Observing System

Key facts

MANAGEMENT:	**NASA Goddard Space Flight Center**
SATELLITES:	**Landsat, *Terra*, the A-Train (*Aqua*, *Aura*, CALIPSO, *Cloudsat*, PARASOL, Orbiting Carbon Observatory), GRACE, SORCE**
FIRST IN ORBIT:	***Landsat 7*, 1999**
LAST IN ORBIT:	**Orbiting Carbon Observatory, scheduled for 2008/2009**

The 1990s brought about a recognition that expanding the methods of observation, combined with a more holistic approach to interpreting the resulting data, offered an opportunity to view the Earth in planetary terms, rather than as the object of isolated forces. These attitudes stemmed from the revelatory images of the Earth provided by photos from space during the Apollo missions. The pictures translated into political action, with the establishment in 1970 of the U.S. Environmental Protection Agency (EPA), and into social awareness, with the advent of annual celebrations of Earth Day. The full technological realization of these and other events resulted in the Earth Observing System, or EOS, during the 1990s. Commissioned to observe and explain global change, the EOS suite came under the management of the Goddard Space Flight Center in 1990. Yet in recognition that climate events happen on a planetary basis, NASA pursued the development and operation of three of the EOS satellites with international partners.

OPPOSITE
Proof of the power of images, the "Blue Marble" photographs of the Earth gave a global perspective to environmentalism during the 1970s. This picture represents a composite from many Earth-observing sources, compiled at the NASA Goddard Space Flight Center.

Tropical Cyclone Favio, imaged on February 22, 2007, by the Moderate-Resolution Imaging Spectroradiometer (MODIS), one of the five instruments aboard the *Terra* satellite. The picture shows Favio crossing the shoreline into Mozambique after losing some of its strength.

The first pair of Cluster satellites being prepared for liftoff aboard a Soyuz launch vehicle in Baikonur, Kazakhstan, July 2000. The Cluster quartet measures the influence of the Sun on the Earth's magnetosphere.

Landsat 7

Key facts

FIRST LANDSAT LAUNCH:	**July 1972**
LANDSAT 7 LAUNCH:	**April 1999**
IMAGES OF EARTH PER DAY:	**250**
LENGTH:	**14 ft / 4.3 m**
DIAMETER:	**9 ft / 2.7 m**
WEIGHT:	**4800 lb / 2180 kg**
ORBIT:	**Circular**
PURPOSE:	**Terrestrial observation**

Landsat embodies the first youth of the new age of environmentalism. Initiated in 1966 and launched in July 1972 as the Earth Resources Technology Satellites (ERTS) program, Landsat satellites represent America's oldest space-based sentinels of the Earth's surface. Because of the project's longevity, it has been subject to a number of institutional adaptations. Renamed Landsat in 1975, the project migrated from NASA, its original home, to NOAA. In 1984, Congress mandated another transfer of Landsat, from the public to the private sector. The following year, Hughes Aircraft and the RCA Corporation joined forces to create the Earth Observation Satellite (EOSAT) Company, offering commercial images of the planet. During the early period of gestation, Landsat underwent successful launches and operations, gradually increasing time aloft. *Landsat 1* flew from 1972 until 1978, *Landsat 2* from 1975 to 1981, *Landsat 3* from 1978 to 1983, *Landsat 4* from 1982 to 1993, and *Landsat 5* from 1984 until the present. During its commercial phase, the satellite did less well. Privately fabricated and controlled, *Landsat 6* failed as a result of a rocket mishap in 1993. During a relatively long transition between NOAA and EOSAT management (1989 to 1992), the funding made available to NOAA became sporadic and unpredictable. Congress finally responded with the Remote Sensing Policy Act of 1992, transferring Landsat once again, this time back to NASA as the lead agency, with NOAA and the United States Geological Survey (USGS) as collaborators.

Landsat became part of the EOS program in 1994. NASA engineers and scientists at Goddard assumed responsibility for the development of the satellite, and contracted with Lockheed Martin for the computer system aboard the spacecraft, and with Raytheon for an instrument known as the Enhanced Thematic Mapper Plus (ETM+). The ETM+ (a derivative of versions flown on *Landsats 4* and *5*) measures solar radiation through eight bandwidths of both infrared and visible radiation. Assisted by a vastly enhanced ground-based data system, *Landsat 7* gathers 250 terrestrial scenes each day, and returns to previously covered areas every sixteen days, repeating the imagery again and again, forming a long-term archive of change. At its launch in April 1999, *Landsat 7* weighed 4800 lb (2180 kg) and measured 14 ft (4.3 m) long and 9 ft (2.7 m) in diameter. It assumed a circular orbit 438 miles (705 km) above the Earth and follows the same ground track as its Landsat predecessors. Each day, ground controllers send commands to the spacecraft directing its daily mission.

ABOVE
A Landsat satellite image, combined with Shuttle Radar Topography Mission data, yielded this perspective view of Malaspina Glacier in southern Alaska.

Landsat contributed to this startling landscape profile of Mount Kilimanjaro, Tanzania.

The *Landsat 7* satellite receives final adjustments in the cleanroom prior to launch.

Landsat records an immensely varied set of observations. One vulcanologist at the University of Hawaii uses Landsat's extraordinarily high-resolution images to plot the movement of new lava flows, which can be distinguished from older flows. He has applied similar techniques to the study of active lava lakes all over the world. A professor of geography at the University of Maryland depends on Landsat to plot the growth of urban sprawl in the Washington, D.C., metropolitan area. He discovered that between 1973 and 1996, the region grew 8½ sq. miles (22 sq. km) per year, and found wide variation in the efficiency of land use in the Maryland and Virginia suburbs. With *Landsat 7*, he will expand his research to cities around the world.

An environmental scientist at the University of Colorado studies the potential—if climate change persists—for dunes and sand sheets presently under grass in the High Plains to be reactivated, causing "dust bowl" conditions to recur. A professor at Michigan State University has harnessed *Landsat 5* data to assess the loss of tropical forests in the Amazon Basin and Southeast Asia during the 1990s; since *Landsat 7*, the amount and detail of the data has increased sharply. A soil scientist with the U.S. Department of Agriculture has combined Landsat observations (which have high definition but cannot penetrate clouds) with radar satellite images (which see through clouds but are less distinct) to create a system of farm monitoring that takes much of the guesswork out of irrigating, fertilizing, seed-crop decision-making, and herbicide applications. A climatologist at NASA Goddard utilizes Landsat to track transformations in the Antarctic ice sheet (concentrating on such features as individual crevasses). Lastly, a researcher with the U.S. Geological Survey relies on Landsat to detect locations and amounts of dry plant life in Yosemite National Park, which greatly assists in the removal of this key component of wildfires.

Labels have been superimposed on the topography revealed by Landsat, here pinpointing McMurdo Station, Antarctica's largest research facility.

Terra

An artist's rendering of the *Terra* satellite in orbit.

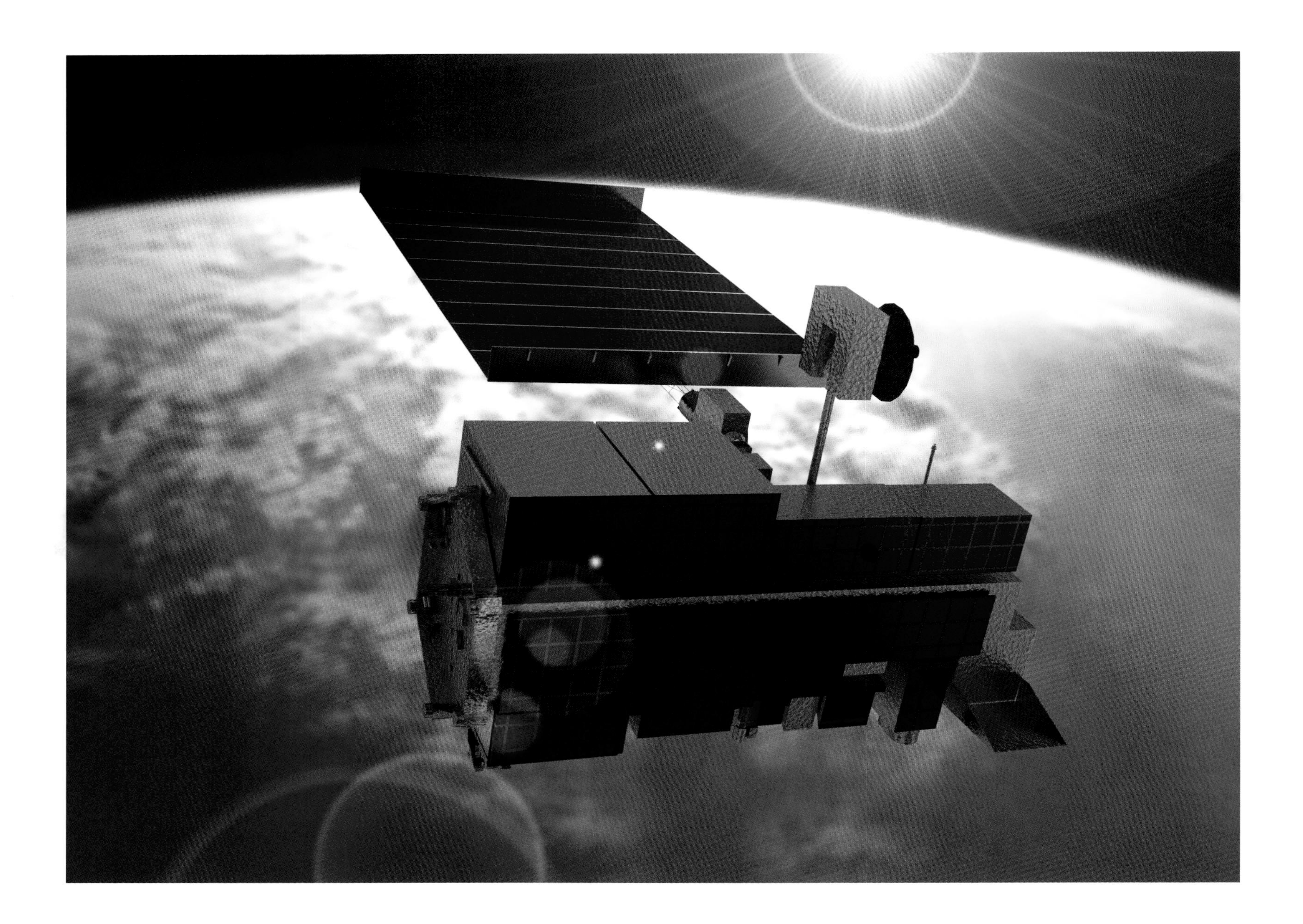

Showing the value of covering the same terrain over time: the Helheim Glacier in eastern Greenland, as seen in 2001, 2003, and 2005. The deterioration of the glacier's margin is obvious.

Human construction shows up plainly in this *Terra* image of palm tree-shaped resorts being built on artificial islands off the Dubai coast, in the Persian Gulf.

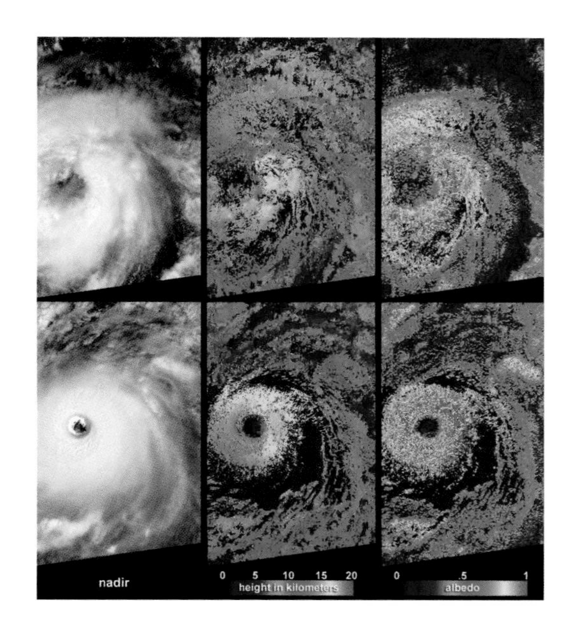

Using the MISR instrument, *Terra* acquired these six images of Hurricane Isabel in September 2003. The top panels persuaded forecasters to categorize it as a hurricane.

Key facts

ROLE:	**EOS Flagship**
LAUNCH:	**December 1999**
INSTRUMENTS:	**ASTER, MODIS, MOPITT, CERES, and MISR**
LIFESPAN:	**15 years expected**
PURPOSE:	**Observe biological processes, climate, natural disasters**

Many refer to *Terra* as the flagship satellite of the EOS family, both because it went into orbit first (except for Landsat, which happened to be part of a pre-existing series), and because of the incredible breadth of its mission. It surveys four landscapes: the atmosphere, the land, the oceans, and the cryosphere (places on the surface of the Earth where water is found in solid form). Launched in December 1999, it will, over its expected fifteen-year lifespan, offer consistent record-keeping and data collection regarding global and seasonal changes in:

- biological processes, including those on the land, in the oceans, on snow and ice, in water vapor and clouds, at surface temperature, and in land cover;
- climate, as altered by human influences;
- natural disasters, such as volcanoes, floods, droughts, and fires.

Terra's five instruments gather data from sunlight and heat radiating from the Earth and send the information to ground stations for processing. These instruments include:

- Advanced Spaceborne Thermal Emission and Reflection Radiometer (ASTER), which consists of three telescope systems capturing *Terra*'s highest-resolution images (in stereo) in the visible range, and in three other wavelengths;
- Moderate-Resolution Imaging Spectroradiometer (MODIS), which observes the entire surface of the planet, day and night, every one to two days, detecting features ranging in size from as small as 820ft to more than 3200 ft (250–975 m). It detects phenomena in thirty-six wavelengths;
- Measurements of Pollution in the Troposphere (MOPITT), a scanning radiometer that detects carbon monoxide and methane concentrations in the lower atmosphere;
- Clouds and the Earth's Radiant Energy System (CERES), two broadband scanning radiometers that detect reflected sunlight, heat radiated from the Earth, and total radiation, which operate night and day;
- Multi-Angle Imaging Spectroradiometer (MISR), a suite of nine cameras, each pointed at a different angle, and each in four wavelengths. The cameras obtain a uniquely multi-dimensional portrait of the planet every nine days, in daylight only.

Among these five instruments, MODIS contributes most broadly, being applicable to all four categories of *Terra*'s scrutiny: atmosphere, land, ocean, and cryosphere. ASTER can be used in all areas but the oceans, and MISR in all but the cryosphere. MOPITT and CERES apply only to the atmosphere.

The range of knowledge derived from *Terra* almost defies belief. Through its instruments, scientists have gained long-term insights about changes in the vegetation that covers the planet; about air pollution sources and patterns; about the mysteries of clouds; about the climate and temperature of the seas; about the flow of heat in the atmosphere; about ice, snow, and floods; about long-range transformations in the ecosystem; about the relationship between urban sprawl and agriculture; about the plant life of the oceans; about methane concentrations and sources; and about airborne particles from volcanoes, dust, industries, and other sources that cool the climate. In all, *Terra* attempts to do nothing less than "give Earth its first physical," as NASA describes it.

EOS takes the A-Train

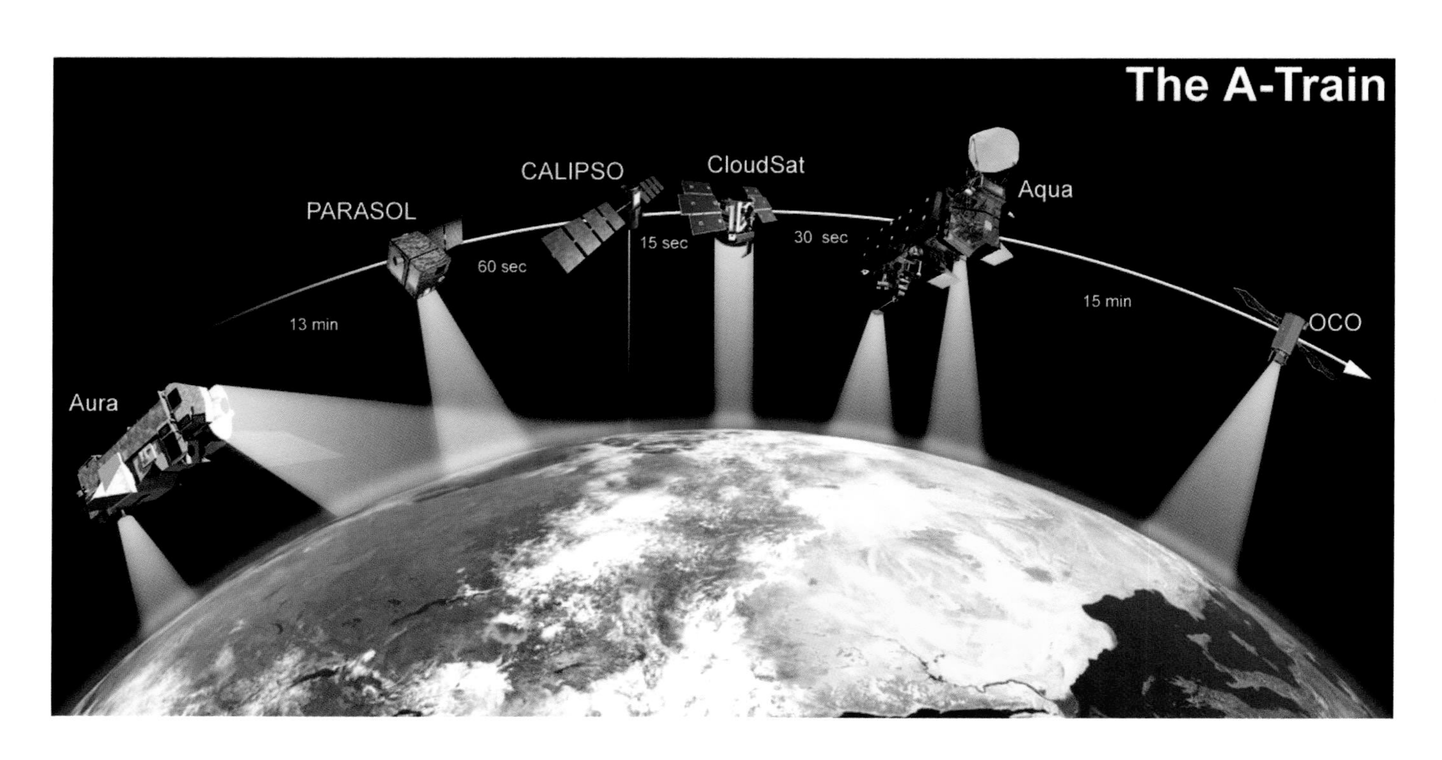

A drawing of the A-Train satellite constellation—part of NASA's Earth Observing System—the components of which work in concert to gather data about climate and factors affecting climate.

The A-Train, as an animator conjectures it might appear if one could stand at one end of the constellation and look ahead to see all six spacecraft at once.

Part of the EOS constellation involves a group of spacecraft called the Afternoon Constellation, shortened to "A-Train." The group derives the name Afternoon Constellation from its track. Flying north—the ascending portion of its orbit—the group crosses the equator at 1.30 pm local time. Its nickname makes reference to the famous Duke Ellington song, "Take the A-Train."

The A-Train consists of six earth-viewing satellites in close formation, flying in Sun-synchronous orbit. The purpose of the A-Train is to capture data about aerosols suspended in the atmosphere, delve into the role of polar stratospheric clouds in the loss of ozone, and study the vertical distribution of ice and water in clouds. The proximity of the satellites to one another enables them to undertake coordinated observations, in addition to their individual missions and capabilities, thereby enhancing the comprehensiveness and accuracy of the overall data. Such split-second teamwork allows the satellites to take multiple "snapshots" in each pass over their constantly evolving subjects.

NASA Goddard Space Flight Center manages the constellation and coordinates activities and data among the six separate missions. Goddard also supervises two of the six projects: the lead satellite, *Aqua*, as well as *Aura*. The Jet Propulsion Laboratory (JPL) in Southern California contributed *CloudSat* to the A-Train. The A-Train also has an important international component: France's Centre National d'Études Spatiales (CNES) joined with NASA Goddard and Langley Research Center to develop and operate the Cloud-Aerosol Lidar and Infrared Pathfinder and Satellite Observation (CALIPSO) satellite, launched together with *CloudSat*. In addition, CNES contributed another satellite to the A-Train known as PARASOL (Polarization and Anisotropy of Reflectances for Atmospheric Sciences coupled with Observations from a Lidar).[1] The key synergies of the overall project will be among *Aqua* (see pp. 70–71) and *Aura* (see pp. 72–73), flying about fifteen minutes apart but on separate orbital tracks; and between CALIPSO and *CloudSat* (see pp. 74–79), launched together and calibrated to adjust to each other's orbit so that complementary data might be taken.

The A-Train constellation is synchronized tightly so that, as the group travels at more than 15,000 mph (24,140 kmph), *Aqua* flies first, *CloudSat* follows between thirty seconds and two minutes later, CALIPSO trails *CloudSat* by no more than fifteen seconds, PARASOL comes one minute later, and *Aura* about thirteen minutes after that. (After the last A-Train satellite, the Orbiting Carbon Observatory (OCO), is launched in late 2008 or early 2009, it will precede *Aqua* by fifteen minutes.)

A still from an animation shows the A-Train viewed perpendicularly as it follows the curvature of the Earth.

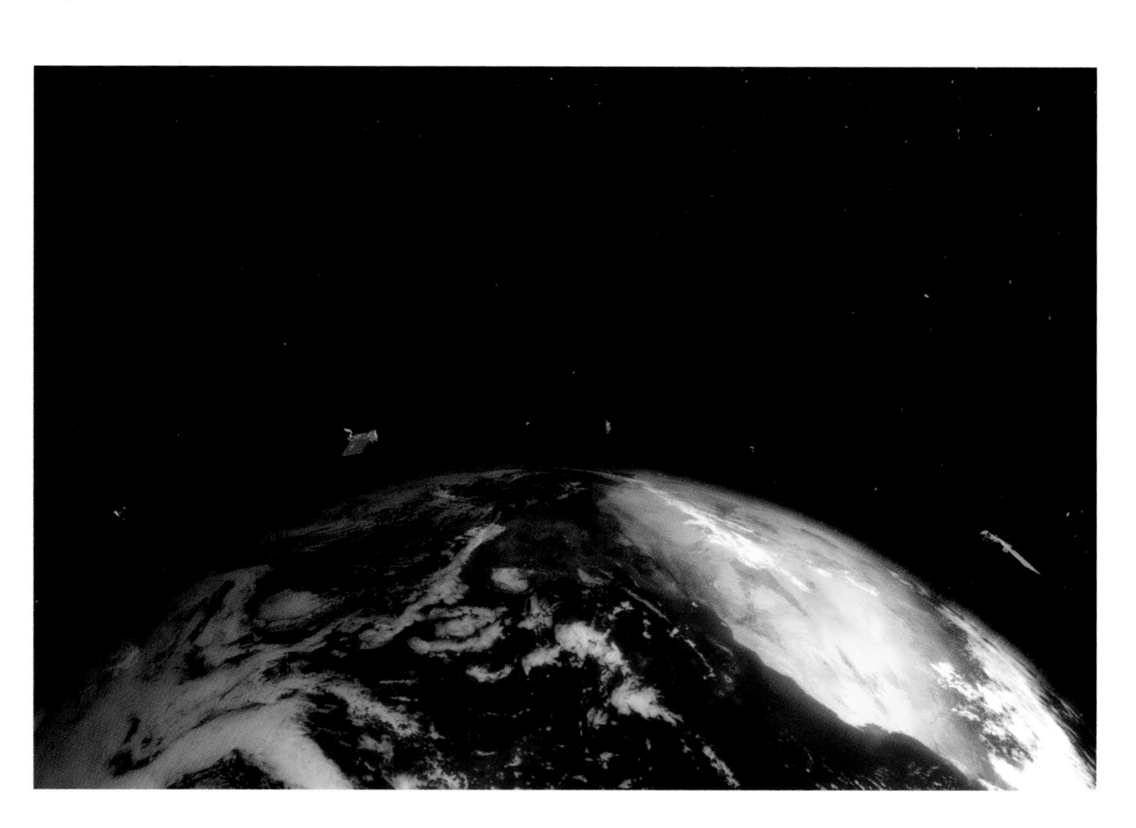

Key facts

MANAGEMENT:	**NASA Goddard Space Flight Center**
NAME:	**Called the "Afternoon Constellation" because it crosses the equator at 1.30 pm local time**
PURPOSE:	**Monitor aerosols in the atmosphere, ozone loss due to polar clouds, distribution of water, and ice in clouds**
FLIGHT CHARACTERISTICS:	**Six satellites fly in tight formation, about thirty minutes separating first from last**

Aqua

Aqua entered Earth orbit on May 4, 2004, the first operational satellite in the A-train. As its name suggests, it specializes in water: the Earth's water cycle, evaporation from the seas, vapor in the atmosphere, rainfall, and so on. On board are six instruments:

- The Atmospheric Infrared Sounder (AIRS) detects temperature profiles, the amount of liquid water in clouds, and the amount of water in the form of rainfall. It also measures the amount of water vapor at different levels in the atmosphere, in conjunction with the AMSU and HSB instruments on the satellite (see below).
- The Advanced Microwave Scanning Radiometer-EOS (AMSR-E) is designed to monitor a variety of water-related atmospheric events, and in particular has helped to improve rain-rate measurement.
- The Advanced Microwave Sounding Unit (AMSU) uses two coordinated sensors to determine temperatures between the Earth's surface and 25 miles (40 km) altitude.
- In common with the same instrument on *Terra* (see pp. 66–67), CERES consists of two identical radiometers that monitor the Earth's reflected sunlight and, by inference, the amount of cloud cover over the planet.
- MODIS, also on the *Terra* satellite, sweeps across the Earth's surface every one to two days, recording the oceanic, land, and atmospheric processes continuously.
- The Humidity Sounder for Brazil (HSB) gives scientists water-vapor readings between the surface and just over 6 miles (9.7 km) altitude.

The *Aqua* satellite undergoes processing before it is sent on its mission to monitor the Earth's water cycle.

The applications of *Aqua* are broad indeed. The satellite has the capacity to improve weather forecasting by upgrading the quality of information used for computer-aided weather reports, reducing the error of initial predictions by a factor of two. Moreover, *Aqua* will facilitate accurate, *worldwide* data collection, a major advance over weather balloons, which can be effective over populous areas, but not over the vastness of the seas. *Aqua* can also take temperature readings as accurate as those from balloon-based sensors.

In addition, this satellite can detect wildfires from the moment they ignite until after they appear to be extinguished. This information is relayed from NASA to the U.S. Forest Service, which relays it, in turn, to firefighters. The MODIS instrument aboard *Aqua* not only pinpoints the blazes, but also tracks their movements. Furthermore, *Aqua* can penetrate thick smoke that deters reconnaissance planes by using infrared, which can likewise assess the intensity of fires. In an entirely different application, MODIS gives *Aqua* the capacity to estimate the extent of snow deposited on land in winter, enabling officials to forecast drought, water conservation, or floods when the snow melts in spring and summer. The influence of the snowpack is life-sustaining in such places as the western United States. It represents about 75 percent of the region's year-round surface water source, so planning is essential.

Aqua has yet another pivotal role involving the planet's climate: to measure the amount of phytoplankton in the oceans. These micro-organisms account for approximately half of the absorption of atmospheric carbon dioxide by the world's plants. In this respect, they have a direct connection to global warming, since carbon dioxide retains heat in the atmosphere. Thanks again to MODIS (this time used in tandem on the *Aqua* and *Terra* satellites), researchers will be able to obtain a global estimate of whether the oceans have been nurturing more phytoplankton or less, based upon changes in the color of the water where these plants flourish.

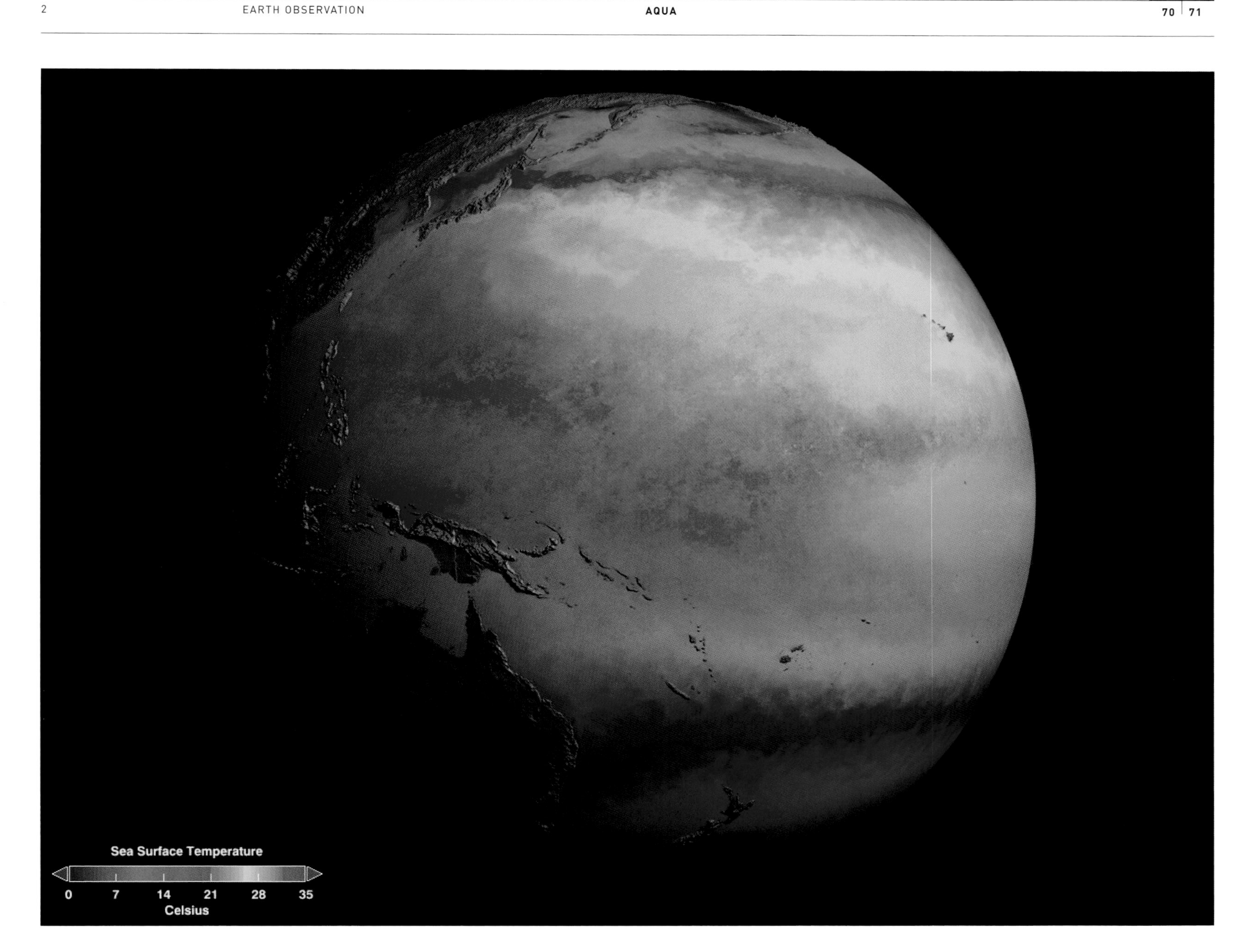

ABOVE
Aqua measures the surface temperature of the world's oceans, here shown graphically over a thirty-two-day period in September-October 2006.

Tropical Cyclone Favio pivots north around the southern tip of Madagascar, as seen by *Aqua* on February 20, 2007.

Key facts

LAUNCHED:	May 2004
PURPOSE:	Chart Earth's water cycle, oceanic evaporation, atmospheric vapor, rainfall
INSTRUMENTS:	AIRS, AMSR-E, AMSU, CERES, MODIS, HSB
SPECIAL CAPABILITY:	Offers worldwide coverage, unlike weather balloons

Aura

A drawing of the *Aura* satellite, seen close up.

Second in the A-Train (by date of launch), the *Aura* satellite went into orbit in July 2004, on a mission to inform scientists and urban officials about the extent and degree of such pollutants as ozone and nitrogen oxide in densely populated environments, with only a few hours' delay in data processing. Ozone absorbs UV light and its depletion has been associated with increases in UV radiation, resulting in crop damage and an increase in cases of skin cancer and eye anomalies. Ozone also occurs as a ground-level air pollutant. In addition to ozone observation, *Aura* will also help authorities forecast climate change.

Aura's researchers hope to answer three pivotal questions based on the data that their satellite returns: Is the ozone layer in the stratosphere recovering? What factors control air quality? How is the climate of the Earth changing? To determine the answers, scientists have put four instruments aboard *Aura*:

- the High Resolution Dynamics Limb Sounder (HIRDLS), an infrared radiometer that detects temperature, gases, and aerosols in the mesosphere, stratosphere, and upper troposphere;
- the Microwave Limb Sounder (MLS), a scanning radiometer that records radiation in the millimeter and sub-millimeter wavelengths;
- the Ozone Monitoring Instrument (OMI), a spectrometer designed to measure reflected and backscattered solar light in the visible and ultraviolet spectra;
- the Tropospheric Emission Spectrometer (TES), an instrument that observes the heat emissions of the Earth's surface and atmosphere in daylight and at nighttime.

Each of the components of *Aura* contributes to a clearer and more complete picture of the extent of human influence on the atmosphere. *Aura* measures climate change, an issue of keen public interest. It observes ozone and water vapor, which are both greenhouse gases; checks ice content in the upper troposphere; and tracks industrial aerosols, dust, smoke, and methane in the troposphere.

Another key area of research is the degree of ozone in the stratosphere. *Aura* records ozone loss in the polar regions in the lower stratosphere during winter, where the largest depletions occur. Although chlorofluorocarbons (CFCs) have been banned around the world, they remain active for long periods and *Aura* measures their transfer from the troposphere to the stratosphere. The satellite also maps broad, planetary ozone change as it sweeps over the Earth and

Key facts

LAUNCHED:	July 2004
PURPOSE:	Observe stratospheric ozone layer, air quality, climate change
INSTRUMENTS:	HIRDLS, MLS, OMI, TES
SPECIAL CAPABILITY:	Measures human influences on the atmosphere

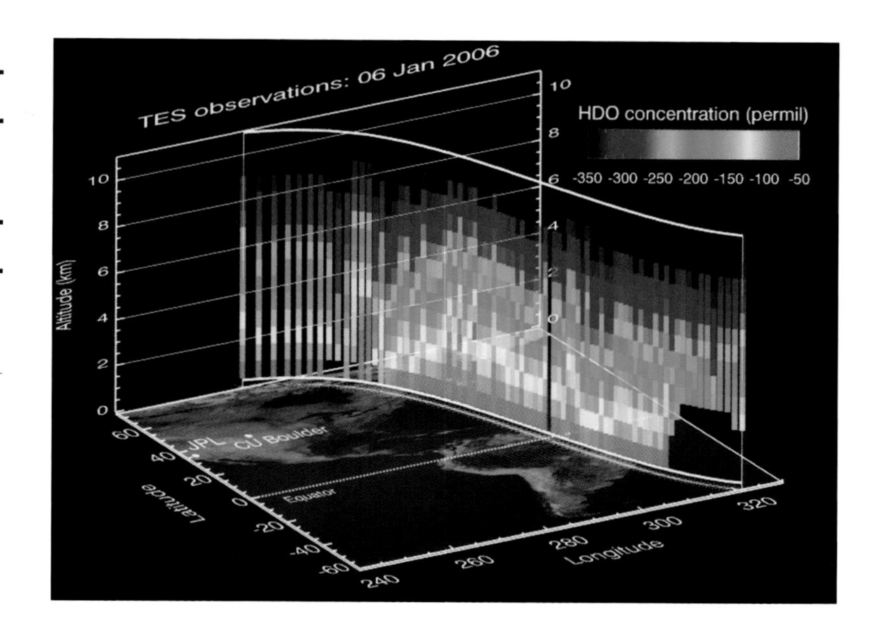

ABOVE, RIGHT
***Aura*'s Tropospheric Emission Spectrometer produced the first direct satellite record of ozone in the lower atmosphere.**

FAR RIGHT
***Aura* takes an ozone measurement of a stratum of the atmosphere. The satellite's Tropospheric Emission Spectrometer enables more precise measurements of ozone, the most elusive of the six criteria pollutants established by the U.S. Environmental Protection Agency.**

RIGHT
***Aura* examines ozone depletion over the poles, illustrated here in three maps of column ozone in the Arctic during the months of January and March 2005 (top and middle), and also on one single day, March 11, 2005 (bottom).**

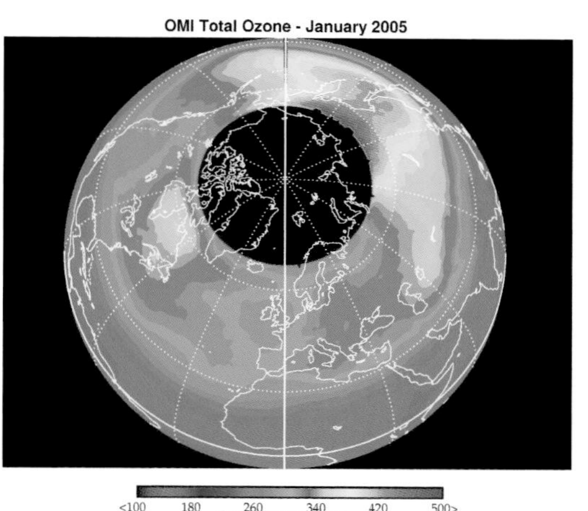

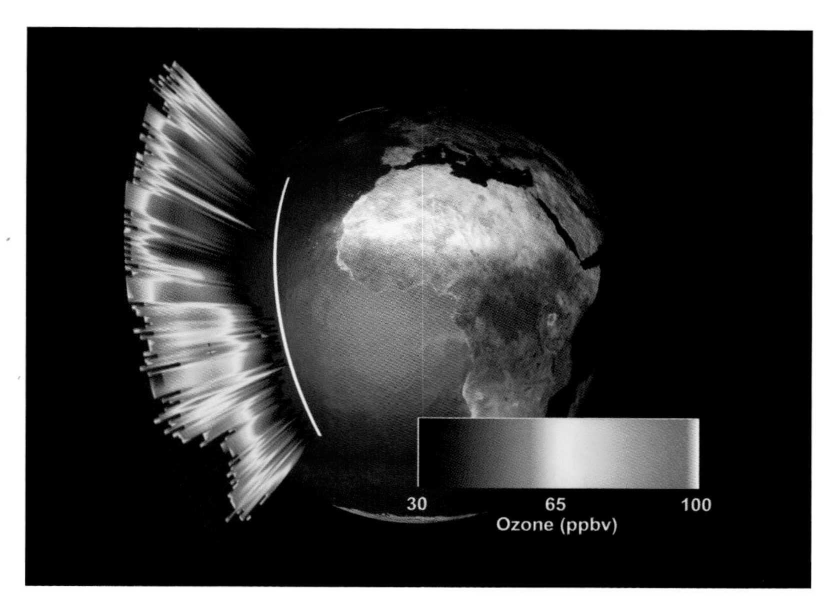

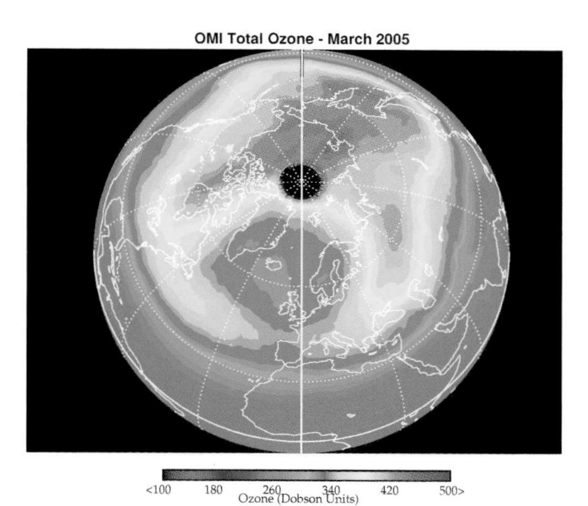

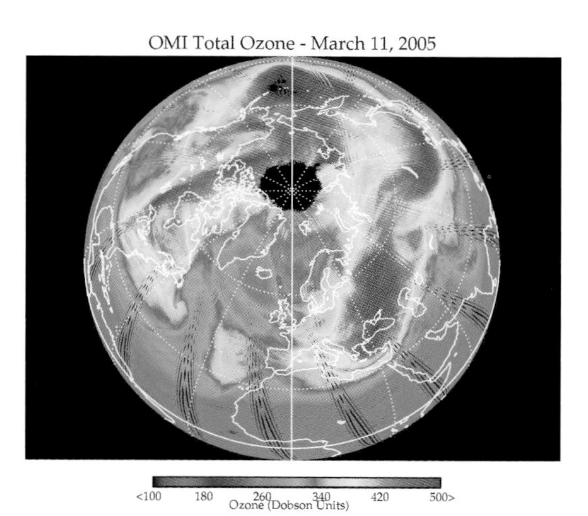

assesses the extent of stratospheric chlorine, which is responsible for depleting ozone.

The issue of air quality is also in *Aura*'s portfolio. Its instruments are able to detect how much air—and pollutants therein—descends from the stratosphere to the troposphere, enabling scientists to separate natural from man-made ozone pollution. *Aura* also measures carbon monoxide and ozone in the upper troposphere (indicators, if present, that a strong vertical movement of pollution has been occurring in the atmosphere), and gives public-health officials accurate readings of ultraviolet radiation reaching the Earth, based upon measurements of clouds and ozone layers.

The twins: CALIPSO and CloudSat

Key facts

LAUNCHED (SIMULTANEOUSLY):	**April 2006**
FLIGHT CHARACTERISTICS:	**Operate together to produce three-dimensional images**
CALIPSO	
MANUFACTURER:	**Alcatel Space (for CNES, the French Space Agency)**
MANAGEMENT:	**NASA Langley Research Center**
INSTRUMENTS:	**CALIOP, wide-field camera, Imaging Infrared Radiometer**
PURPOSE:	**Measure aerosols in the atmosphere (both human and natural in origin)**
CloudSat	
MANUFACTURER:	**Ball Aerospace and Technologies**
MANAGEMENT:	**NASA Jet Propulsion Laboratory, Pasadena, California, U.S.**
INSTRUMENT:	**Cloud Profiling Radar**
PURPOSE:	**Monitor cloud structure, rainfall, snowfall, and rain from condensation**

An artist's view of CALIPSO in orbit. CALIPSO uses laser light, among other techniques, to observe clouds and aerosols.

An artist's conception of the CALIPSO satellite (foreground) paired in flight with *CloudSat*.

Two of the A-Train satellites flew into orbit together. They lifted off aboard a two-stage Boeing Delta II rocket from Complex 2 at Vandenberg Air Force Base, California, on April 28, 2006. At the top of the 128-ft (39-m) stack, a composite payload fairing protected these two A-Train satellites: just under the nose cone, CALIPSO, and below it, *CloudSat*. Operating together, this pair provides a three-dimensional picture of clouds and airborne particles as they regulate the world's climate and air quality.

An international collaboration between NASA and the French CNES, the CALIPSO satellite arrived in California from the Alcatel Space Facility at Cannes, France, in May 2005. CNES also provided one of the satellite's instruments and integrated the payload with the spacecraft, and now contributes spacecraft mission operations. The Institut Pierre Simon Laplace in Paris oversees the radiometer science, as well as data validation and archiving. For NASA, Langley Research Center leads the mission with systems engineering and payload operations, as well as processing and storing data, while Goddard retains overall project management.

CALIPSO is equipped with three instruments: the Cloud-Aerosol Lidar with Orthogonal Polarization (CALIOP), a wide-field camera, and an Imaging Infrared Radiometer (IIR, provided by CNES). CALIOP detects the intensity of light reflected by aerosols and clouds using the instrument's Lidar, which behaves like radar, with the difference that the radio waves used in radar are supplanted by light waves emitted by a laser. CALIPSO discharges a brief but very intense burst of laser light over a wide area, and the returning light allows a three-dimensional analysis of a vertical profile (about 330 ft/100 m wide) of the atmosphere for aerosols, as well as for thin, water-bearing, and ice clouds. The wide-field camera captures digital images of the area surrounding the vertical "slice" taken by Lidar, enabling researchers to know the

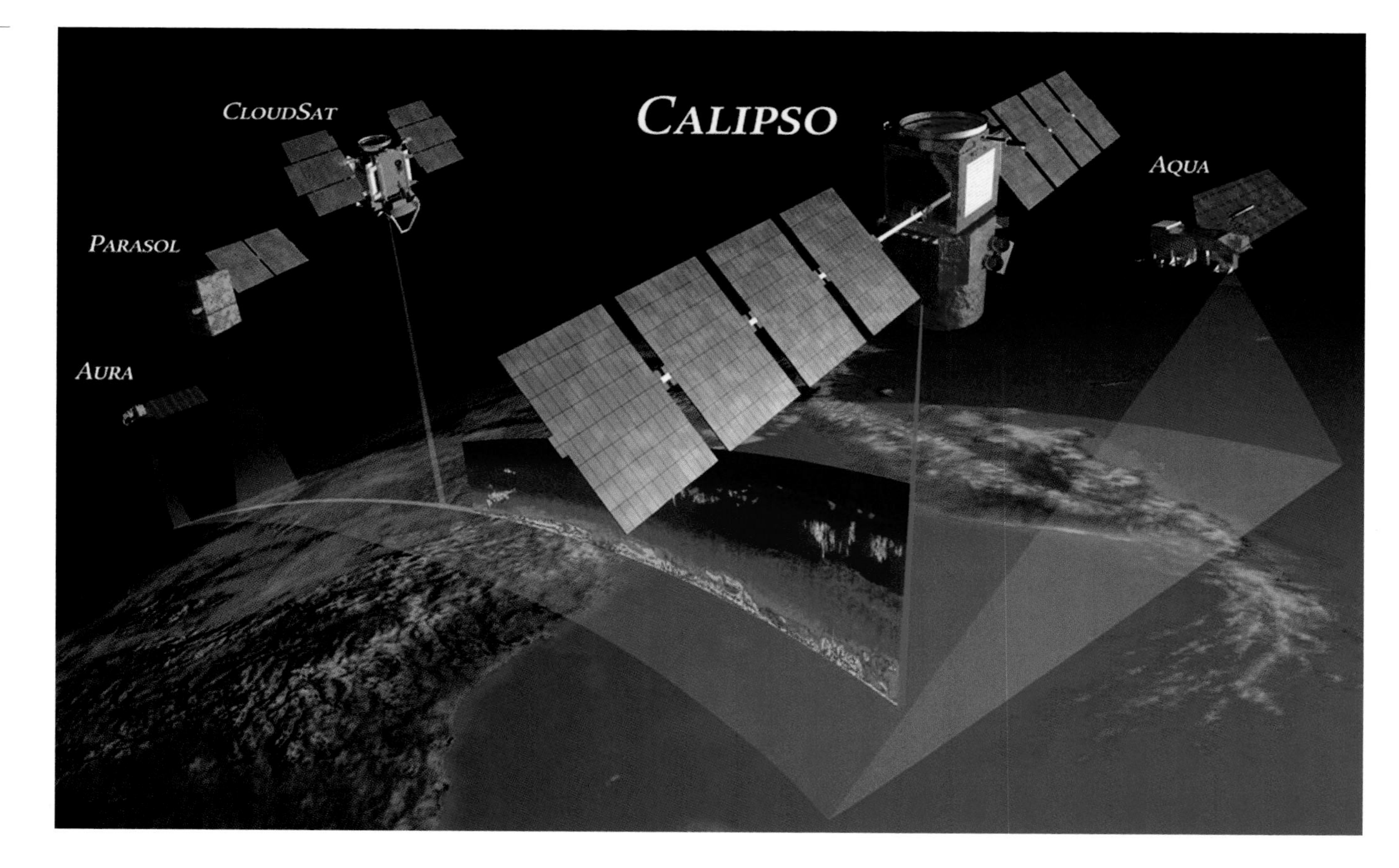

A drawing of CALIPSO in relation to the other A-Train satellites. (In order to emphasize CALIPSO in this image, the artist has reversed the true positions of *CloudSat* and CALIPSO.)

context of their atmospheric sample. The IIR, set at three thermal infrared wavelengths, works in synchronization with CALIOP to optimize measurements of cirrus cloud emissions and particle size.

CALIPSO's scientists and engineers conceived and built it to shed light on important, but as yet missing, data regarding aerosols, a diverse collection of airborne particles from such natural sources as volcanic eruption, desert dust, sea salt, and smoke from forest fires; and human pollutants derived from fossil fuels, coal, oil, and chemicals unleashed by industry. Missing so far is a knowledge of the outer limits of the altitude of such aerosols. Those lower in the atmosphere may be washed down by rain, but those far above can travel great distances and affect far-away countries. CALIPSO will also bring greater certainty about the height of clouds, an important factor in accurate precipitation prediction. Ultimately, CALIPSO will offer insights into the formation of clouds and aerosols and how they influence the planet's weather and air quality.

Like CALIPSO, *CloudSat* has an international component, although not a full partnership such as that between CNES and NASA. JPL managed the project, and also developed the satellite's radar, to which the Canadian Space Agency made hardware contributions. Other partners include the U.S. Air Force, which provided ground operations and communications, and Ball Aerospace and Technologies, designer and fabricator of *CloudSat* under contract to JPL.

The scientists and engineers involved in *CloudSat* wanted a satellite capable not only of dissecting the interior anatomy of clouds, but also of revealing their distribution in order to comprehend more clearly their relationship to the Earth's radiation budget and its hydrological cycles. Clouds offer one of the biggest challenges to the pursuit of accurate weather models. The novelty, and also the familiarity, of *CloudSat* rests with its lone instrument, the Cloud Profiling Radar (CPR). The first of its kind to fly in space, the CPR design and subsystems bear a strong resemblance to the airborne cloud radar that has operated aboard NASA's DC-8 flying laboratory since 1998. One thousand times more sensitive than standard weather radars, CPR uses millimeter wavelengths, as opposed to the centimeter wavelengths of ground-based radars. Centimeter wavelength radars can detect raindrop-sized particles; millimeter wavelength radars will pick up the more minute particles of liquid water and ice, the building blocks of the cloud masses that govern weather.

RIGHT
A portion of CALIPSO undergoes inspection in the cleanroom.

FAR RIGHT
The CALIPSO satellite stands upright in a processing facility.

OPPOSITE
A drawing of *CloudSat* in orbit, as it gauges the vertical structure of the clouds below for rain and snow potential.

NASA
CALIPSO
CLOUDSAT
BOEING
BOEING
NASA

OPPOSITE
***CloudSat* and CALIPSO take off together in April 2006 aboard a Delta II rocket, from Vandenberg Air Force Base, California.**

BELOW
***CloudSat* has the capacity to view storm systems in profile. Here it details (in three strip images and a key) the structure of Hurricane Daniel as it evolved over five days in July 2006. These strip readings were paired with the top three images captured by a GOES satellite.**

BOTTOM, LEFT
More sensitive than any weather radar to date, the equipment on *CloudSat* will enable researchers to determine the influence of clouds on the distribution of the Sun's energy in the atmosphere.

BOTTOM, RIGHT
An artist illustrates the manner in which *CloudSat* concentrates microwave energy to sense cloud particles, ice, and water.

CloudSat offers a number of benefits as a result of its CPR and its orbital vantage point. It allows global surveys of the percentage of clouds that yield rain; vertically oriented estimates about the degree of water and ice in clouds; the capacity to detect snowfall, from space; and data about the efficiency with which the atmosphere produces rain from condensated materials. By penetrating the structure, makeup, and full vertical extent of clouds, scientists expect more accurate weather (short-term) and climate (long-term) forecasts. Until then, guesswork prevails about the net influence of different types of clouds on the climate. For instance, what is the overall heating and cooling balance between the high, thin cirrus clouds (which allow the passage of sunshine on to the planet and check the escape of solar radiation), versus the thick, low cumulus clouds (which block incoming sunlight and reflect it back into space)? For now, gaps in our knowledge of clouds hinder prediction about climate change. But by comprehending such factors as cloud height and structure, the extent of cloud cover, and the amounts of water and ice therein, the scientific community should be able to anticipate with more certainty the direction in which climate change may be headed. Such predictions alone would constitute a worthwhile outcome of A-Train research.

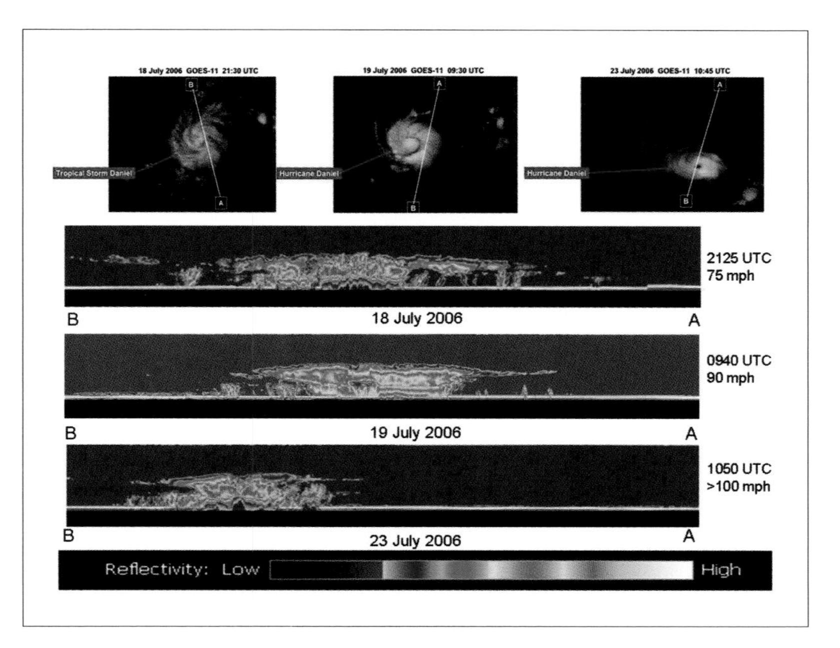

Gravity Recovery and Climate Experiment

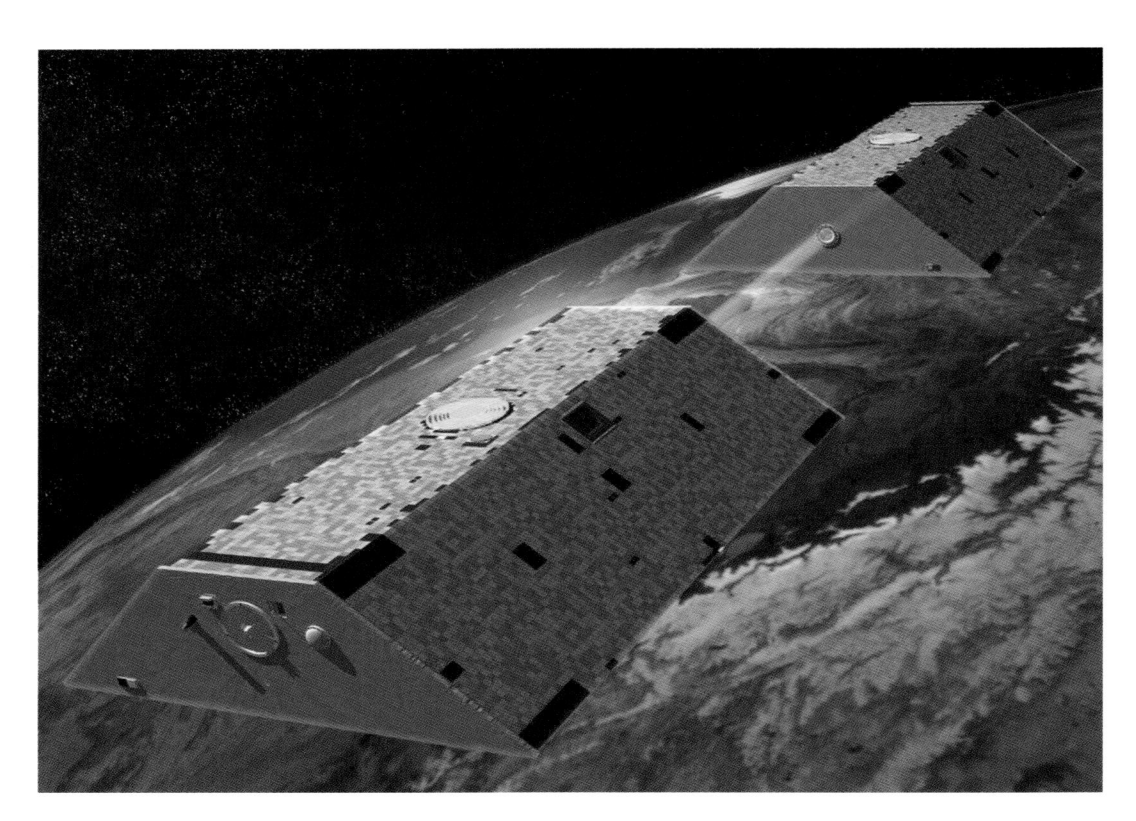

Key facts

LAUNCHED:	**March 2002**
FLIGHT CHARACTERISTICS:	**Twin satellites flying in the same orbit, 137 miles / 220 km apart**
ORBIT:	**Nearly polar**
MANAGEMENT:	**DLR and NASA**
SPECIAL CAPABILITY:	**The GRACE satellites act as science instruments, rather than as passive platforms**
INSTRUMENTS:	**HAIRS, Superstar Accelerometers, Star Tracker Camera, GPS**
PURPOSE:	**Detect big Earth-based climate factors, such as oceanic currents and aquifers**

Two illustrations of the GRACE satellites: a hypothetical close-up showing the satellites' features (above), and how they might look in orbit. Variations in their relative positions give clues about inconsistencies in the Earth's gravitational field.

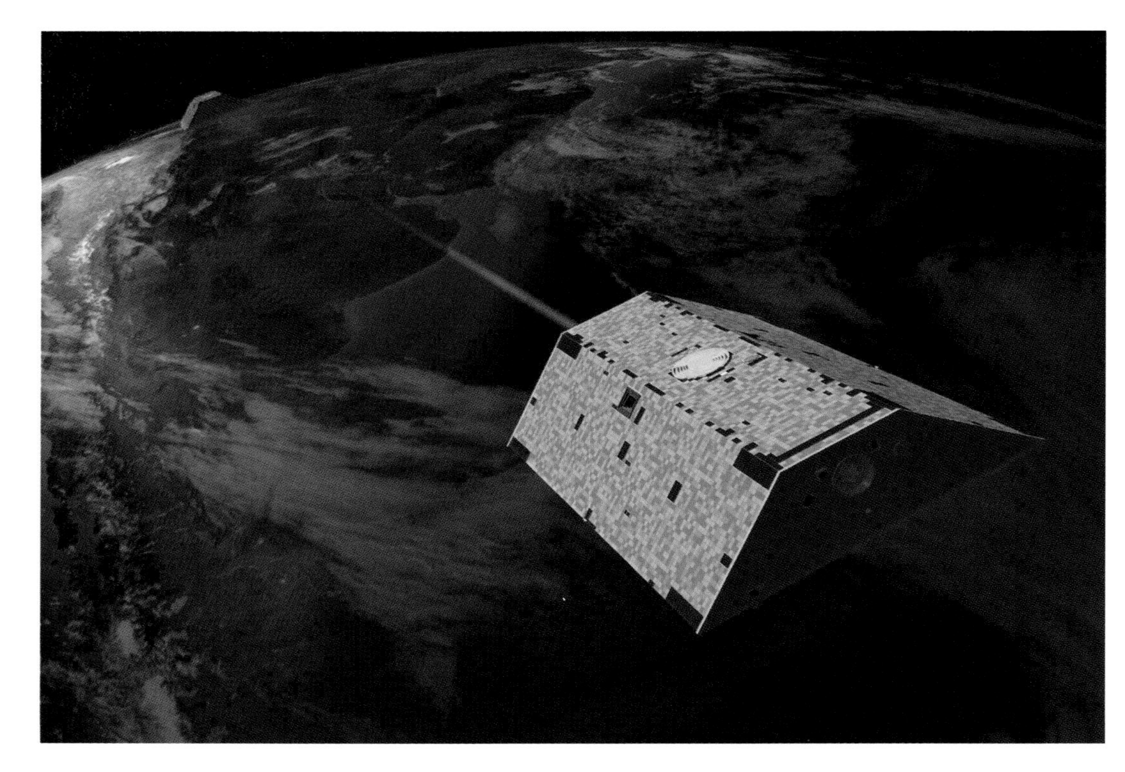

The Gravity Recovery and Climate Experiment (GRACE) represents yet another satellite in the Earth Observing System, though it is not part of the A-Train. Actually, GRACE consists of twin satellites, orbiting 310 miles (499 km) above the Earth, almost over the poles, sixteen times a day. One flies ahead of the other by about 137 miles (220 km), roughly the distance between San Diego and Los Angeles.

GRACE not only relies on two vehicles, but also has twin parents: the Deutsches Zentrum für Luft- und Raumfahrt (the German Aerospace Center, known as the DLR) and NASA. The Germans obtained launch services from Eurorockot, a partnership between Russia's Khrunichev State Research and Production Space Center and Daimler-Chrysler Aerospace. The satellites lifted out of the atmosphere from Russia's northern spaceport at the Plesetsk Cosmodrome aboard a three-stage Rockot launch vehicle. Perhaps the GRACE team hoped for luck by taking off on St. Patrick's Day (March 17), 2002.

Unlike most other satellites, the GRACE twins operate as the main science instruments themselves, rather than as passive platforms. They were created to measure the Earth's gravitational field in order to discern the actions of vast planetary water systems that affect climate, and their designers needed to consider the satellites' mass, balance and stability, dimensional stability, and aerodynamics. By monitoring the relative distance between each satellite, scientists hoped to determine precise inconsistencies in the Earth's gravitational force, which are themselves the result of variations in the distribution of the planet's mass due in part to the movement of water. Four instruments accomplished this task:

- The High Accuracy Intersatellite Ranging System (HAIRS) detects changes in the relative position of the ends of the satellites. The system factors into its readings changes in relative position caused by non-gravitational factors as well.
- The Superstar Accelerometers on each satellite measure non-gravitational phenomena that cause variation in the satellites' traveling speed, such as atmospheric drag, solar radiation, and control-thruster forces. This information corrects the HAIRS system.
- The Star Tracker Camera provides attitude control, as well as data that make it possible for accelerometer readings to be translated into inertial references.
- The Global Positioning System (GPS) sends data to the onboard computers that synchronize the two separate telemetry streams of the satellites, allowing determination of their relative positions to within roughly $\frac{3}{4}$ in. (1.9 cm), if not better.

GRACE's highly accurate reading of the Earth's gravitational field, in addition to the production of monthly maps showing its variations, differentiates these satellites from all previous missions. With them, scientists can make informed inferences about the biggest climate-related mechanisms of the planet, such as ocean currents (deep and shallow), ice melting, stored water and evapotranspiration, and underground aquifers.

Because the flow of water affects the planet's gravitational field, the GRACE readings enable hydrologists to study the motions of water across the globe, with a degree of detail not previously possible. Because water migrates relatively quickly, and because it is found beneath the surface of the Earth (in aquifers), on it (as solid and liquid), and above it (as vapor), it has been difficult to track its movements comprehensively. GRACE uses gravity as a hunter uses a hound, tracing the variations in the gravitational field to determine the sources, pathways, and collecting points of the world's largest, yet most elusive, agent. Resource planners have benefited greatly from this knowledge, which has enabled them to predict storage and, as a result, to distribute water more wisely among such competing consumers as agriculture, industry, and municipalities.

GRACE also offers insights into the impact of global warming. Changes in the mass of the polar caps and Greenland register with GRACE as minute deviations in gravitational force; teamed with aircraft observations of glacier height and the altimeter measurements of other satellites, this information will enable scientists to state the probability of whether the world's ice sheets are growing or shrinking. Moreover, the unique capabilities of GRACE will also help answer the vexing question of whether the rise in sea levels results from the warming of the oceans and thermal expansion, or from water melting from the glaciers. GRACE data will make it possible to discover whether the seas have been rising not just because of oceanic warming or ice meltdown, but also as a result of "post-glacial rebound." This phenomenon occurs when the Earth's crust adjusts itself vertically in response to the reduced burden of weight caused by the recession of the last ice age, resulting in higher sea levels at the coasts. Similarly, GRACE will facilitate more accurate comprehension of the currents and temperatures of the seas by factoring in the effects of gravity on these phenomena.

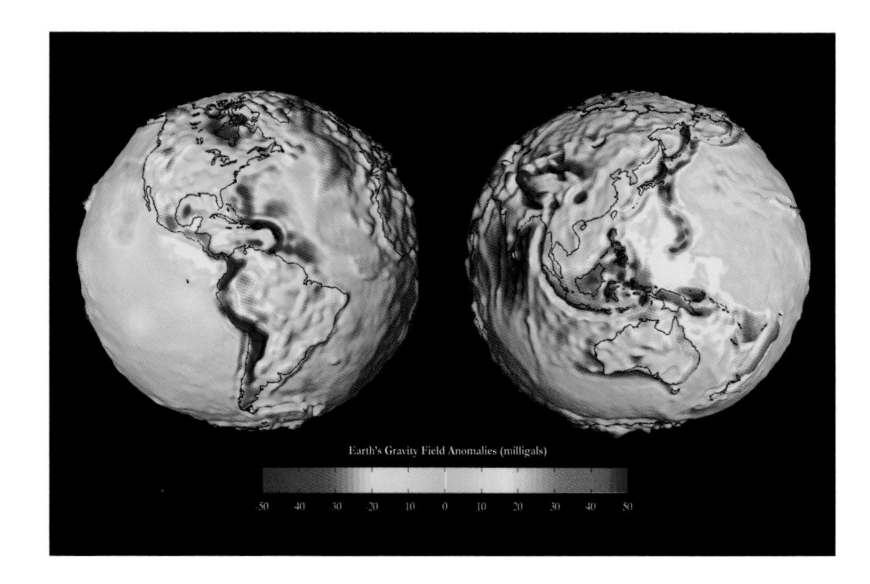

ABOVE
This is perhaps the most accurate map yet produced of the Earth's long wavelength gravity field. Fluctuations in gravitational force may suggest subtle movements in the planet's climate-related mechanisms.

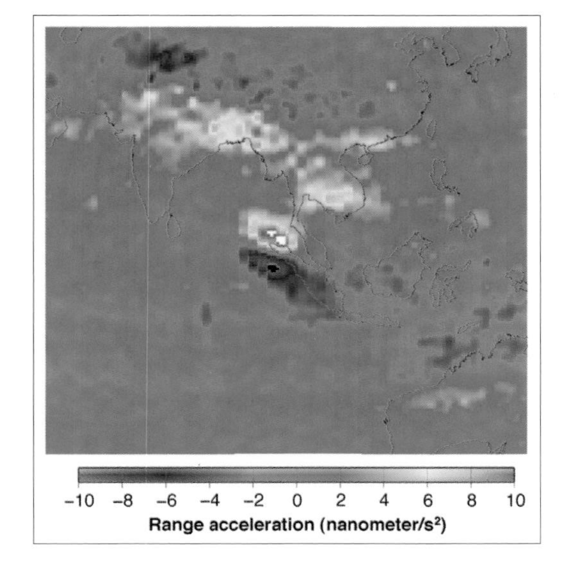

A two-dimensional map of the Earth's gravity field, based on data collected by GRACE.

Solar Radiation and Climate Experiment

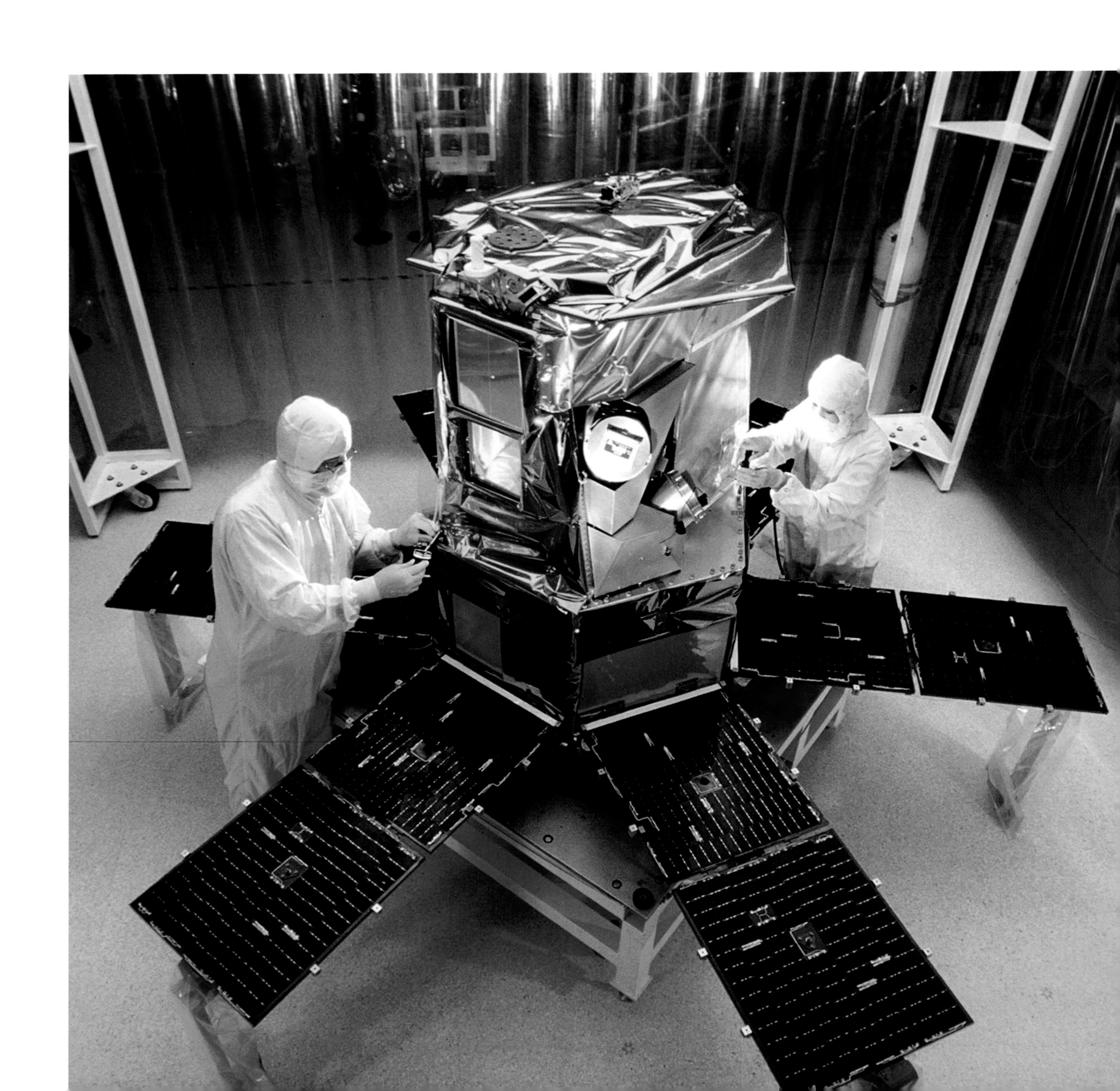

NASA's Earth Orbiting System constellation has also turned its eye on the Sun. The small Solar Radiation and Climate Experiment satellite (SORCE), launched aboard an aircraft-borne Pegasus XL rocket on January 25, 2003, has provided NASA with precise data about the Sun's radiation in order to determine the influence of solar radiation on the Earth's atmosphere and climate. The Laboratory for Atmospheric and Space Physics at the University of Colorado, Boulder, operates the mission under the management oversight of Goddard Space Flight Center.

The University of Colorado team also conceived and built SORCE's four instruments:

- The Total Irradiance Monitor (TIM) measures the total amount of solar radiation at the upper reaches of the atmosphere, and does so four times every day. A precision aperture behind one of TIM's shutters locates the exact location over which it collects the sunlight, measures the sunlight as heat, and then determines the ratio of the power of radiation to the area exposed to it, yielding a precise Total Solar Irradiance (TSI) reading.
- The Spectral Irradiance Monitor (SIM) gives scientists the opportunity to learn how the solar cycles influence near-infrared and visible wavelengths. Combined with observations from other satellites, it will help reveal how much light reaches the land and oceans, and how much remains in the lower portion of the atmosphere, resulting in heating. Based on new prism spectrometer technology that covers a broad range of light, SIM can capture 90 percent of the TSI, breaking down the light into wavelengths and directing them to five photodiodes that measure each one. Like TIM, SIM takes four measurements a day.
- Two Solar Stellar Irradiance Comparison Experiment (SOLSTICE) instruments measure ultraviolet in order to calibrate SORCE as a whole. Like an earlier SOLSTICE launched on the Upper Atmospheric Research Satellite in 1991, these instruments train their attention on eighteen bright-blue stars, measuring the stars' ultraviolet emissions. The ultraviolet radiation from these stars remains constant, so if SOLSTICE detects variations, scientists know they must correct their data accordingly.
- The XUV Photometer System (XPS) measures energy emitted by the Sun's hot and variable corona in the form of solar X-ray. The heating and ionization of the upper atmosphere results primarily from the burst of photons emitted by solar flares. Observations of this kind began in the 1960s, with sounding rockets and some of the earliest satellites, but the infrequency of the data collection rendered the results fragmentary. Starting in the 1990s, a continuous flow of data has occurred, which SORCE contributes to with greater accuracy than previously.

As SORCE captures information about the Sun's influence on the Earth, new knowledge comes to light about the planet's energy balance. Scientists have long wanted to know how much solar energy lights the Earth, and what happens to it once inside the atmosphere. It seems that about 30 percent of the energy directed at the Earth is reflected back into space by clouds, sand, aerosols, and even rooftops. The lower atmosphere absorbs 20 to 25 percent, leaving 45 to 50 percent of visible light to penetrate the land and oceans. SORCE's mission also concerns itself with changes in the Sun itself. Because the amount of solar energy emitted varies daily due to sunspots, scientists want to know if there are predictable patterns of such activity. They know that eleven-year cycles exist, but have also learned that variations occur over the centuries. For instance, between 1650 and 1715, astronomers saw no sunspots on the Sun, followed by a gradual rise in such incidents. In addition, SORCE offers clues about global warming. As much as half of the 1°F (0.6°C) increase in global temperature over the last century may be the result of increased sunspots, indicating that as much of the rise might be caused by the Sun as by increased carbon dioxide emissions. Finally, SORCE may offer improved predictability of solar changes and a better understanding of the Sun's impact on climate change by determining the average solar irradiance emitted over a decade.

OPPOSITE
The SORCE satellite in the cleanroom, being prepared for launch.

An artist's rendition of SORCE in orbit. SORCE measures solar radiation in the upper reaches of the Earth's atmosphere.

Key facts

LAUNCHED:	**January 2003**
LAUNCHER:	**Aircraft-borne Pegasus XL rocket**
MANAGEMENT:	**NASA Goddard Space Flight Center**
INSTRUMENTS:	**TIM, SIM, SOLSTICE, XUV Photometer System**
PURPOSE:	**Measure solar radiation to learn the Sun's impact on Earth's atmosphere and climate**

Geostationary Operational Environmental Satellites

If the Earth Observing System's network of satellites suggests long-term trends to scientists about the health of the Earth as an environmental system, the long-serving Geostationary Operational Environmental Satellites (GOES)—which orbit in one position relative to the rotation of the Earth unless they are moved intentionally—focus on the vagaries of weather. Thus, programmatically separate from EOS, the GOES mission operates in the here and now, offering current weather data and maps to forecasters and meteorologists around the world rather than long-term predictions.

The GOES family of satellites has operated for more than thirty years and number thirteen to date. *GOES 1* through *7* entered orbit aboard the Delta launch vehicle, *8* through *10* flew the Atlas–Centaur combination, and *11* and *12* lifted off on the Atlas II booster. Over time, they became more capable, more complex, and heavier (*GOES 1* to *7* ranged from 650 to 874 lb/295 to 396 kg; with *GOES 8*, the bulk rose to 4600 lb/2080 kg, hence the heavier Atlas–Centaur launcher). The first in the series, *GOES 1*, began service in October 1975, placed (like all of its successors) in geosynchronous orbit, in this case at the equator over the Indian Ocean. This satellite and the following seven in the GOES series (launched between June 1977 and April 1994) have all been decommissioned. Until spring 2006, only *GOES 9* through *12* remained operational, having been sent aloft between May 1995 and July 2001. *GOES 9* has been leased to Japan and flies over the mid-Pacific. *GOES 10* covers the eastern Pacific, and *GOES 11* orbits in storage mode (ready to replace those that fail). *GOES 12*, over the Amazon, serves the meteorological needs of most of the United States.

Like all of the GOES family since 1983, *GOES 13*—sent into orbit aboard a Delta IV rocket on May 14, 2006—reflects a formal, working partnership between NOAA and NASA. NOAA decides on technical requirements, determines the need for new spacecraft, distributes the meteorological data, operates the satellites, and supplies funding. Representing NASA, the Goddard Space Flight Center designs and procures the satellites, conducts verification testing, and does the initial on-orbit checkout.

GOES 13 through *15* (called the G series, and referred to as *GOES-N*, *-O*, and *-P* until they go into orbit) constitute an improved version of the existing constellation. Built by Boeing Space and Intelligence Systems, *GOES 13* and its cohorts weigh 6908 lb (3133 kg) at launch and measure 13 ft 9 in. (4.2 m) in length and 6 ft 2 in. (2 m) in width. *GOES 13* to *15* surpass the earlier spacecraft in a number of respects. Among many advantages, they have ten- rather than seven-year life spans, offer command data rates of 2000 bytes per second (compared to 250 in the earlier versions), and can pinpoint the source of severe weather events down to about 4922 ft (1500 m), halving the current 9844 ft (3000 m).

The four main instruments aboard *GOES 13* enable weather forecasters to receive timely meteorological data about the size, intensity, and direction of storms. The GOES spacecraft work together in pairs high above the equator, gathering data over about 60 percent of the Earth's surface. Their instruments consist of:

- an imager (or imaging radiometer) that produces continuous images of the planet's surface, including storm developments; the oceans; cloud temperature, cover, and height; water vapor; and surface temperature;
- a sounder that measures emitted radiation in one visible and eighteen thermal bands, enabling it to detect temperature, moisture, ozone, and reflected solar radiation;
- a Space Environment Monitor (SEM) that measures energetic particles (protons, electrons, and alpha particles), the extent and direction of the Earth's geomagnetic field, and solar X-ray emissions;
- a solar X-ray imager that observes solar X-rays in order to detect flares on the Sun.

Such data help forecasters determine which solar phenomena will influence the Earth's weather and communications.

Besides the invaluable services to humanity provided by the GOES satellites, they also serve as a global search-and-rescue system, operated by NOAA. The GOES satellites detect signals beamed at 406 MHz, the band at which emergency transmitters on aircraft, ships, and individuals (carrying Personal Locator Beacons) broadcast. The satellites relay the alert signals to the U.S. Mission Control Center in Suitland, Maryland, which in turn transmits the distress data to appropriate Rescue Coordination Centers around the U.S. and in twenty-six other countries. Since the end of 2004, as many as 18,000 persons may have been rescued around the world as a result of GOES.

Key facts

MANAGEMENT:	**NASA Goddard Space Flight Center and NOAA**
MANUFACTURER:	**Boeing Space and Intelligence Systems**
GOES 1 SATELLITE:	**Began service in October 1975**
GOES 13 SATELLITE:	**Launched May 2006**
INSTRUMENTS:	**Imaging radiometer, sounder, Space Environment Monitor, Solar X-ray Imager**
PURPOSE:	**Monitor the size, intensity, and direction of storms for forecasters**

BELOW
A top-down image of Hurricane Ileana over the eastern Pacific in August 2006, generated by GOES. Color has been used to indicate temperature.

RIGHT
A Delta IV rocket lifts off from Cape Canaveral Air Station, Florida, in May 2006, carrying aloft the *GOES 13* satellite.

BELOW, RIGHT
An artist's conception of the *GOES 13* satellite in orbit, tracking the weather from a geosynchronous orbit.

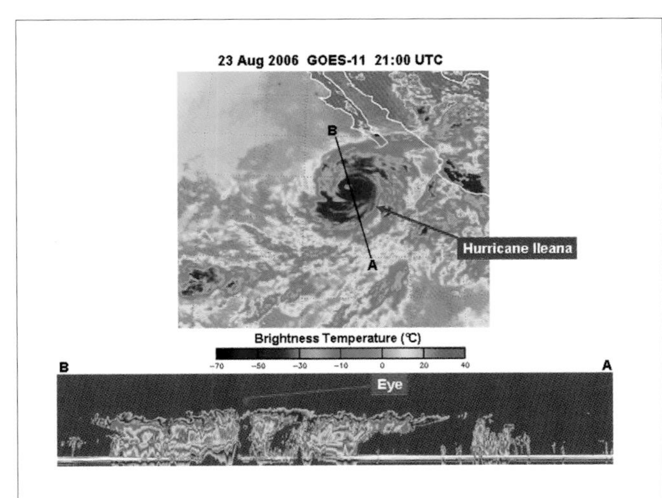

Jason-1

Key facts

MANAGEMENT:	**CNES and NASA Jet Propulsion Laboratory**
PREDECESSOR:	**TOPEX / Poseidon satellite, also under CNES / NASA management**
LAUNCHED:	**December 2001**
FLIGHT CHARACTERISTICS:	**TOPEX / Poseidon and *Jason-1* operated in tandem**
INSTRUMENTS:	**Poseidon-2 Altimeter, DORIS, Jason Microwave Radiometer, Laser Retroreflector Array, Blackjack GPS**
PURPOSE:	**Observe surface oceanic activity**

An artist's rendering of TOPEX/Poseidon monitoring the currents and circulation of the oceans.

Like the collaboration involved in CALIPSO (see pp. 74–76) a partnership existed between the French Space Agency (CNES) and NASA regarding measurements of the surface of the world's oceans. From the late 1960s, engineers and scientists speculated about the benefits of studying the oceans through space-borne radar altimeters, realized in part in Seasat (1978) and Geosat (1985–89). The Franco-American alliance on this technology had its origins at JPL, where researchers first conceived of The Ocean Topography Experiment (TOPEX), a satellite altimeter designed to plot the ocean's contours. Meanwhile, scientists at CNES planned for a complementary mission called Poseidon, after the Greek god of the seas. The two agencies joined forces in 1987, and in August 1992 the TOPEX/Poseidon spacecraft went into orbit aboard an Ariane rocket, launched from ESA's Space Center in Kourou, French Guiana.

For the next thirteen-and-a-half years, TOPEX/Poseidon stood sentinel over 95 percent of the Earth's ice-free seas, charting the patterns of oceanic circulation and observing how heat migrated in the oceans, a key ingredient of climate change. It enabled scientists to see for the first time major, long-term trends in oceanic circulation at its surface. For instance, in 1997 TOPEX/Poseidon provided a global perspective to El Niño and La Niña (temporary changes in the climate of the Pacific Ocean), allowing a more comprehensive knowledge of the origins of short-term weather patterns such as these. It also mapped the world's tides, a feat never accomplished before, and measured sea levels to about 2 in. (5 cm) accuracy.

A year after TOPEX/Poseidon went aloft, a seminar in St. Malo, France, incubated the concept of a second-generation TOPEX/Poseidon satellite called *Jason-1* (a reference to Jason, leader of the Argonauts in the Greek myth). It flew into orbit on a Delta rocket from Vandenberg Air Force Base, California, in December 2001. (Between that date and TOPEX/Poseidon's demise—roughly four years later—the two spacecraft operated in tandem, allowing for double the coverage of the seas as well as the calibration of each other's instruments.) As with TOPEX/Poseidon, ground-control responsibilities for *Jason-1* are shared between the French and American space agencies. A command center in Toulouse, France, sends directions to *Jason-1*'s instruments on board and processes and coordinates spacecraft control and data reception. NASA's JPL operates the American-made instruments, monitors the spacecraft, beams commands to it, and generates real-time data products. The spacecraft weighs about 1103 lb (500 kg) when fueled, is almost 10 ft (3 m) tall, and covers the Earth every ten days.

Jason-1 has five main instruments. The French Space Agency contributed the Poseidon-2 Altimeter, the main onboard instrument that measures the altitude of the spacecraft over the sea, wind speed, and the height of waves to within about ¾ in. (1.9 cm). CNES also provided the Doppler Orbitography and Radiopositioning Integrated by Satellite (DORIS), to determine the orbital path and to make ionospheric corrections for its Poseidon-2 satellite. NASA, meanwhile, added the Jason Microwave Radiometer (JMR), which records the water-vapor content along the trajectory of the altimeter signal in order to make corrections, as well as the Laser Retroreflector Array (LRRA), which calibrates altimeter readings and coordinates with ground stations to track the orbit. NASA also fitted *Jason-1* with a BlackJack Global Positioning receiver that receives signals from other satellites and ground stations in order to make precise orbital determinations.

Weather forecasters, climatologists, maritime enterprises, and industry all benefit from the observations of *Jason-1*. Mariners equipped with its altimeter-based readings on wave height, wind speed, and currents can reduce sailing time. Data about fundamental oceanic activity over extended periods inform those attempting to predict global climate. Information about tidal movements may have a decisive impact on political decisions involving coastal land use, along which half of the world's population lives. Offshore oil drillers also want to know about currents and eddies in remote parts of the seas as their platforms push farther from the shorelines and into deeper waters.

Jason-1's creators expect a long lifespan, not for this satellite in particular—its mission was planned to run for three years—but for a succession of spacecraft that will follow it. Next in line will be *Jason-2*, based upon an agreement between CNES, NASA, NOAA, and the European Organisation for the Exploitation of Meteorological Satellites (EUMETSAT). A launch in 2008 has been announced.

***Jason-1*, shown here in a drawing, was developed jointly by the French and American space agencies as the successor to TOPEX/Poseidon.**

SAC-C

TOP
An artist's conception of *SAC-C* in orbit, a product of CONAE and NASA.

ABOVE
The *SAC-C* satellite lifts off from Vandenberg Air Force Base, California, in November 2000.

In order to comprehend another key part of the Earth's ecosystem, NASA and the Argentinian National Space Activities Commission (CONAE) agreed to collaborate on a spacecraft called *SAC-C* (*Satelite de Aplicaciones Cientificas-C*). It involved broad international cooperation. CONAE developed the spacecraft itself, as well as several of its key instruments. Goddard Space Flight Center guided the overall project management, JPL contributed two instruments, and NASA offered the launch vehicle. In addition, the Italian Space Agency (ASI) supplied the solar panels and two GPS receivers; the Danish Space Research Institute (DSRI) provided the Magnetic Mapping payload; the French CNES put aboard an experiment to test electronic circuitry subjected to space radiation; and the Brazilian Space Agency (AEB) made facilities available for system-level testing. (An earlier partnership forged between NASA and CONAE involved a satellite known as *SAC-B*, which failed to separate from its Pegasus launch vehicle.)

When *SAC-C* lifted off from Vandenberg Air Force Base in California in November 2000, aboard a Delta rocket, it did so in tandem with a NASA satellite called *EO 1* (*Earth Observing 1*), although the two had no operational connection. *SAC-C* weighed 1045 lb (474 kg) and measured approximately 7 x 6 x 5 ft (2.1 x 1.8 x 1.5 m), with solar arrays 10 ft (3 m) across. The launch of *SAC-C* marked an important milestone for Argentina. The media publicized the event, quoting a NASA official who described it as "another step in Argentina's emergence as a spacefaring nation with a truly sophisticated Earth-observing satellite."[2] Just as in the early days of Sputnik and Explorer, a recognized achievement in space confers a special status on the countries that succeed, well out of proportion to the hard data actually realized, however important. It remains a relatively select club.

The hard data in the case of *SAC-C* involves the imaging of terrestrial and coastal landscapes. In addition, it measures aspects of the dynamics and structure of our planet's geomagnetic field, ionosphere, and atmosphere. Specifically, *SAC-C*'s scientists want to learn about the influence of high-energy radiation on advanced electronics; discover more about the Earth's magnetic field and related interactions with the Sun; sample atmospheric phenomena related to weather and climate change; and monitor the condition of the world's biosphere and environment on land and sea. Paradoxically, *SAC-C* also turns its eye on a relatively small detail of life on Earth: the migration patterns of the Franca whale. Also known as the southern right whale, these creatures frequent the coast of Argentina, where they raise their calves in winter and spring. The Franca almost became extinct after unchecked hunting in the eighteenth and nineteenth centuries, but after fifty years of protection in the twentieth, may now number between 4000 and 5000. The satellite will look for clues in their migrations that may further hasten their increase.

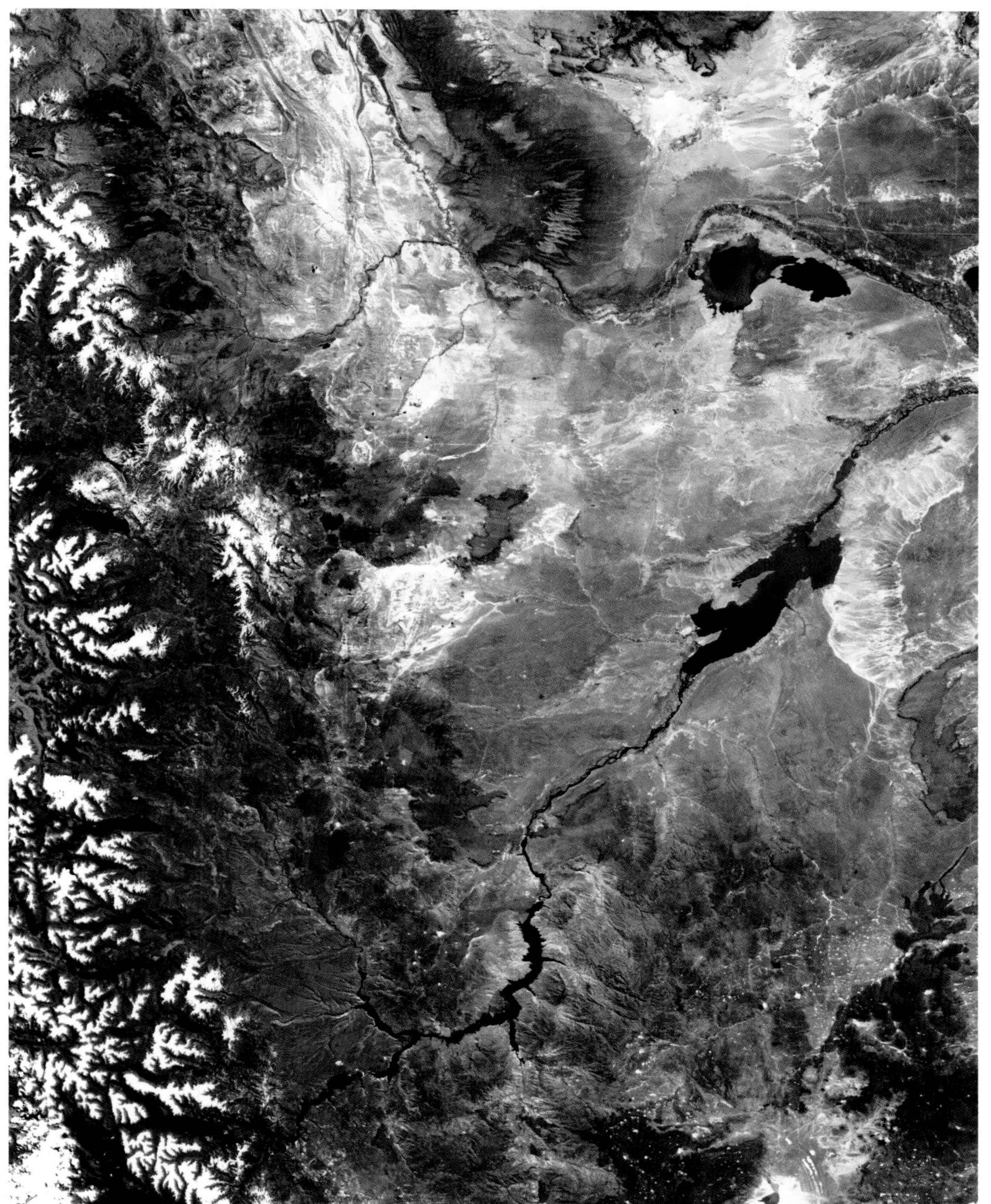

Key facts

MANAGEMENT:	**CONAE and NASA Goddard Space Flight Center**
LAUNCHED:	**November 2000**
PURPOSE:	**Measure impact of high-energy radiation on electronics, as well as weather and climate change, the world's biosphere, and Earth's magnetic field**
INSTRUMENTS:	**MMRS, HRTC, High Sensitivity Camera, Italian Star Tracker, Data Collection System, Digital Transponder, MMP, GOLPE, SHM**

***SAC-C* makes images of terrestrial and coastal landscapes. This example shows the flow of lakes and rivers in relation to the topography.**

SAC-C will accomplish these feats through a suite of eleven instruments:

- Multispectral Medium Resolution Scanner (MMRS), conceived and fabricated by CONAE, to examine the marine and terrestrial environments of Argentina, including droughts, deforestation, diseased habitats, floods, and even parasites;
- High Resolution Technological Camera (HRTC), also from CONAE, to corroborate some of the MMRS observations as it, too, concentrates on Argentina;
- High Sensitivity Camera (HSC), an instrument provided by CONAE for forest fire and electrical storm detection;
- Influence of Space Radiation on Advanced Components (ICARE) experiment, the product of CNES, to measure the impact of high-energy particles and radiation on cutting-edge electronic equipment;
- Italian Star Tracker (IST), to test orbit determination and the satellite's fully autonomous system;
- Whale Tracker Experiment, operational when *SAC-C* flies over the South Atlantic Ocean;
- Data Collection System, to gather environmental data from inexpensive ground platforms;
- Digital Transponder, to facilitate amateur radio communications;
- Magnetic Mapping Payload (MMP), designed in Denmark to plot the Earth's electromagnetic field continuously for one year;
- GPS Occultation and Passive Reflection Experiment (GOLPE), an instrument conceived at JPL that uses an advanced GPS system connected to four high-gain antennas. GOLPE measures the Earth's gravity field and also employs a new GPS remote-sensing capability to detect short-term weather, and long-term climate, change;
- Scalar Helium Magnetometer (SHM), another JPL product, which complements the MMP instruments, adding hardware and electronics.

Tropical Rainfall Measuring Mission

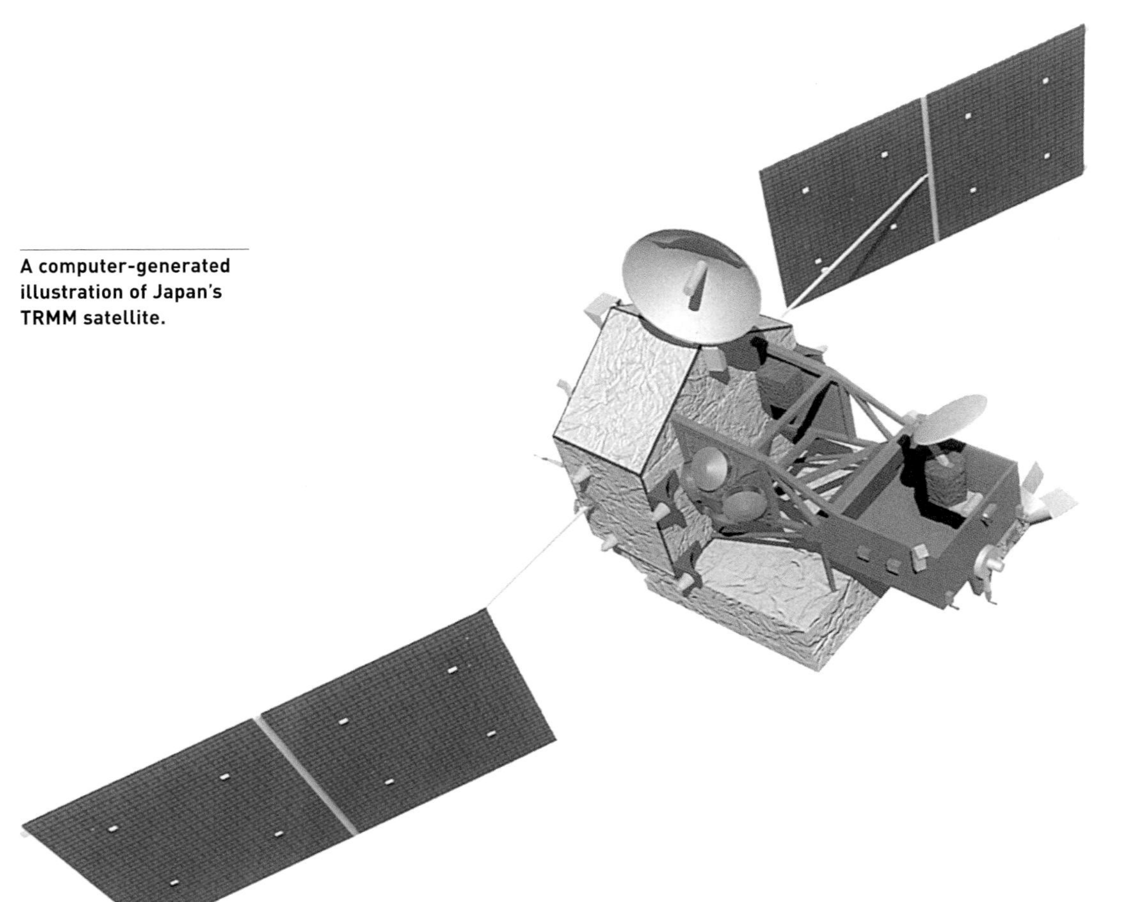

A computer-generated illustration of Japan's TRMM satellite.

Key facts

LAUNCHED:	November 1997
MANAGEMENT:	JAXA and NASA Goddard Space Flight Center
PURPOSE:	Monitor rainfall in the tropics and its influence on global climate
INSTRUMENTS:	Precipitation Radar, TRMM Microwave Imager, Visible Infrared Scanner
FLIGHT CHARACTERISTICS:	Views the Earth only from 36°N to 36°S

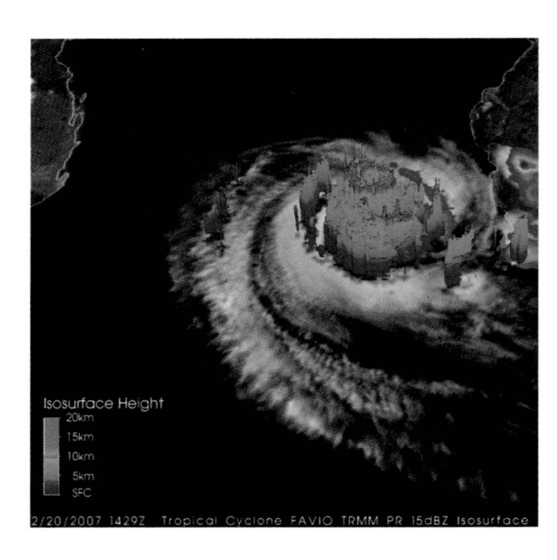

A view of Hurricane Favio captured in February 2007 by TRMM, a joint project between JAXA and NASA.

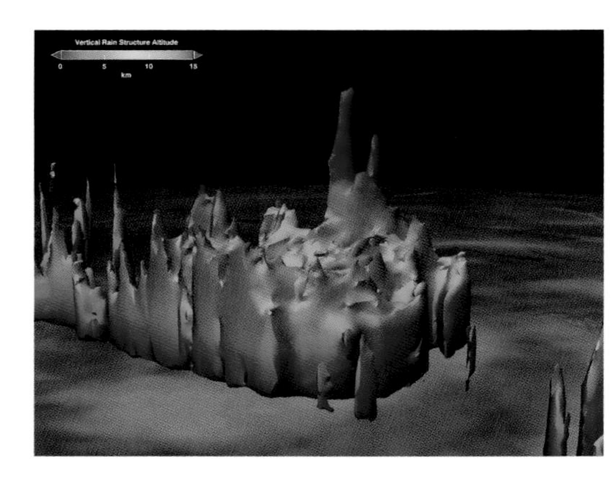

A representation, generated by TRMM in October 2005, of the vertical rain structure in the eyewall of Tropical Storm Wilma.

TRMM records Tropical Storm Wilma in October 2005 as it gathers strength in the Caribbean Sea.

In November 1997, the Japanese National Space Development Agency (forerunner of the present-day Japan Aerospace Exploration Agency, or JAXA) launched the Tropical Rainfall Measuring Mission (TRMM) satellite from its Tanegashima Space Center. JAXA's partner in the project, NASA, expressed an interest in terminating the mission in 2004, but protests kept TRMM aloft. Ten years after it began, it remains in orbit, where it will stay until at least September 2009, unless it runs out of fuel first.

This "gray fox" of the Earth observation world fulfills a pivotal function. More than two-thirds of the Earth's rain falls in the tropics, a phenomenon that renders them the main distributor of heat on the planet and, as such, a key engine of climate formation. Just as *Jason-1* dissects the actions of the oceans to discern their role in climate change, TRMM watches the tropics to learn their contribution to global climatology. The TRMM mission originated with a NASA–Japan collaboration, in which the Japanese lofted the satellite aboard an H-II rocket and contributed a unique Precipitation Radar, the main and most innovative instrument on the satellite. For NASA, the Goddard Space Flight Center provided the spacecraft, two other primary instruments, and operational control.

The mission—initially expected to last only three years—concentrated on four objectives: to collect and study tropical and subtropical rainfall over an extended period; to learn the interactions of sea, air, and land that result in changes in rainfall and climate; to improve the modeling of tropical precipitation processes in order to achieve better prediction; and to test and upgrade satellite measurement techniques. TRMM fills a vast gap in the collection of data, monitoring open country, where ground measurement remains scarce, and scanning the oceans, where even fewer Earth-based measurement exists. Moreover, depending on its readings, TRMM tests, verifies, and modifies the existing computer models of the Earth's climate system.

On board its 7744-lb (3513-kg) structure, orbiting at roughly 217 miles (349 km) altitude, TRMM carries three main instruments devoted to rain measurement:

- Precipitation Radar (PR): By measuring the radar reflectivity of the cloud systems, PR became the first instrument in space capable of mapping the three-dimensional structure of storms, yielding data on rain type, distribution, and intensity, as well as storm depth.
- TRMM Microwave Imager (TMI), a multi-channel microwave radiometer, measures cloud water, water vapor, and rainfall in the atmosphere, more accurately over the seas than over land.
- Visible Infrared Scanner (VIRS) detects radiance in five bandwidths, from infrared to visible, enabling rough estimates of precipitation.

Two other less prominent instruments contribute to TRMM's effective sentry work. The Lightning Image Sensor (LIS), an optical telescope, captures the release of atmospheric and air-to-ground lightning discharges, with relevance to storm dynamics and the release of latent heat. The CERES instrument acts as a sensor to detect visible to infrared energy rising from the atmosphere and from the planet itself, as opposed to energy received from the Sun. The relationship between the two constitutes the Earth's radiation budget.

As TRMM's time in orbit inevitably runs down, talks have taken place about replacing the world's only satellite devoted exclusively to rain measurement, and the only one with the Precipitation Radar. To expand TRMM's coverage, NASA, JAXA, ESA, CNES, the Canadian Space Agency, and the Indian Space Agency have discussed a nine-satellite constellation, each fitted with a microwave radiometer and two precipitation radars. These potential partners see a benefit in expanding the work of TRMM, which currently views the Earth only from 36°N to 36°S, samples rainfall only once a day, and cannot detect light rain or frozen precipitation.

The TRMM satellite undergoes inspection in the cleanroom.

Cluster

ABOVE
A drawing of all four of the Cluster spacecraft, as they might look in orbit.

LEFT
ESA's Cluster mission patch, showing a graphic of the solar wind and the four satellites.

Key facts

MANAGEMENT:	**ESA**
MANUFACTURER:	**Daimler Benz Aerospace**
LAUNCHED:	**July 2000 (first pair) and August 2000 (second pair)**
FLIGHT CHARACTERISTICS:	**Four satellites flying in tight formation**
PURPOSE:	**Observe the Earth's magnetosphere**
INSTRUMENTS:	**FGM, ASPOC, EDI, STAFF, WHISPER, Wide Band Data Instrument, PEACE, CIS, RAPID**
ORBIT:	**Highly elliptical**

Somewhat like the American A-Train satellites, the European Space Agency's four Cluster spacecraft fly together in a prescribed array, in this case in the shape of a pyramid. Cluster differs, however, in that it varies its pattern according to the phenomena it observes, and also orbits in a highly elliptical circle around the Earth.

The four identical satellites needed to be produced and assembled with the greatest possible care to ensure uniformity, no trivial chore for the ESA team. The project originated in late 1982 with a call for proposals, became an authorized mission in 1986, and sat ready, atop an Ariane-5 booster, for liftoff in June 1996. All four of the Cluster spacecraft got a free ride aboard the Ariane because this mission was the rocket's first test flight. Like many free offers, it turned out to be expensive. The star-crossed launch lasted just thirty-seven seconds into the mission, when high aerodynamic loads initiated the booster's auto-destruct mechanism, destroying the entire program in a single ball of flame.

Following introspection about whether to recreate the whole constellation again, ESA resurrected the program with the objective of launching in 2000, at the peak of the solar cycle. Because of budget constraints, another trip on the Ariane had to be ruled out. Instead, the team turned to Russia and the reliable Soyuz rocket, combined with the Russian Fregat upper stage. Daimler Benz Aerospace (predecessor

RIGHT
An artist's perspective on the flying Cluster quartet. Together, these spacecraft attempt to assess the effects of the collision between the Earth's magnetosphere and the solar wind.

FAR RIGHT
An artist's rendering of one of the Cluster pairs just after separation in orbit.

An artist's concept of the four Cluster satellites being buffeted by waves of magnetic and electrical energy as they fly through the Earth's magnetotail. The rust-colored satellite to the left of the Cluster quadruplets is ESA's *Double Star TC-1* spacecraft.

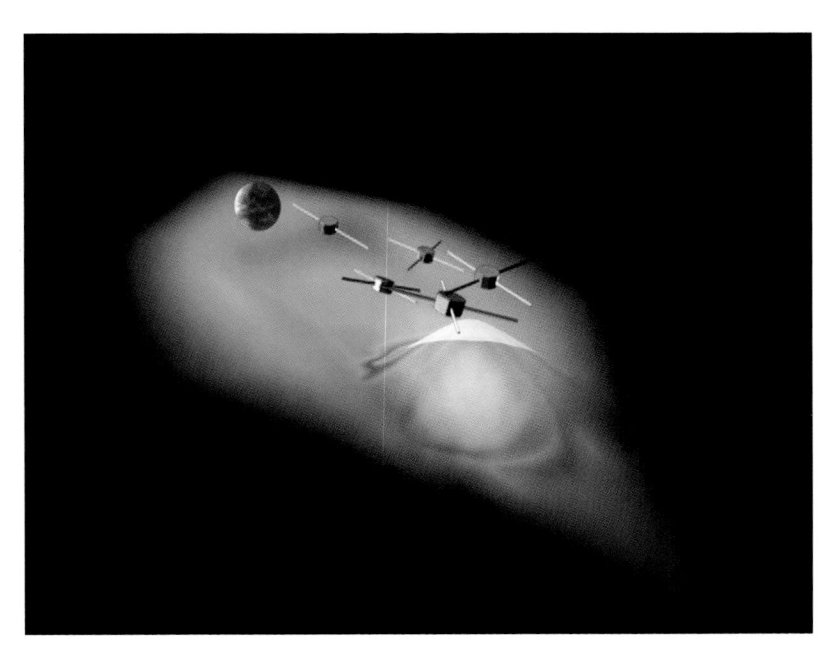

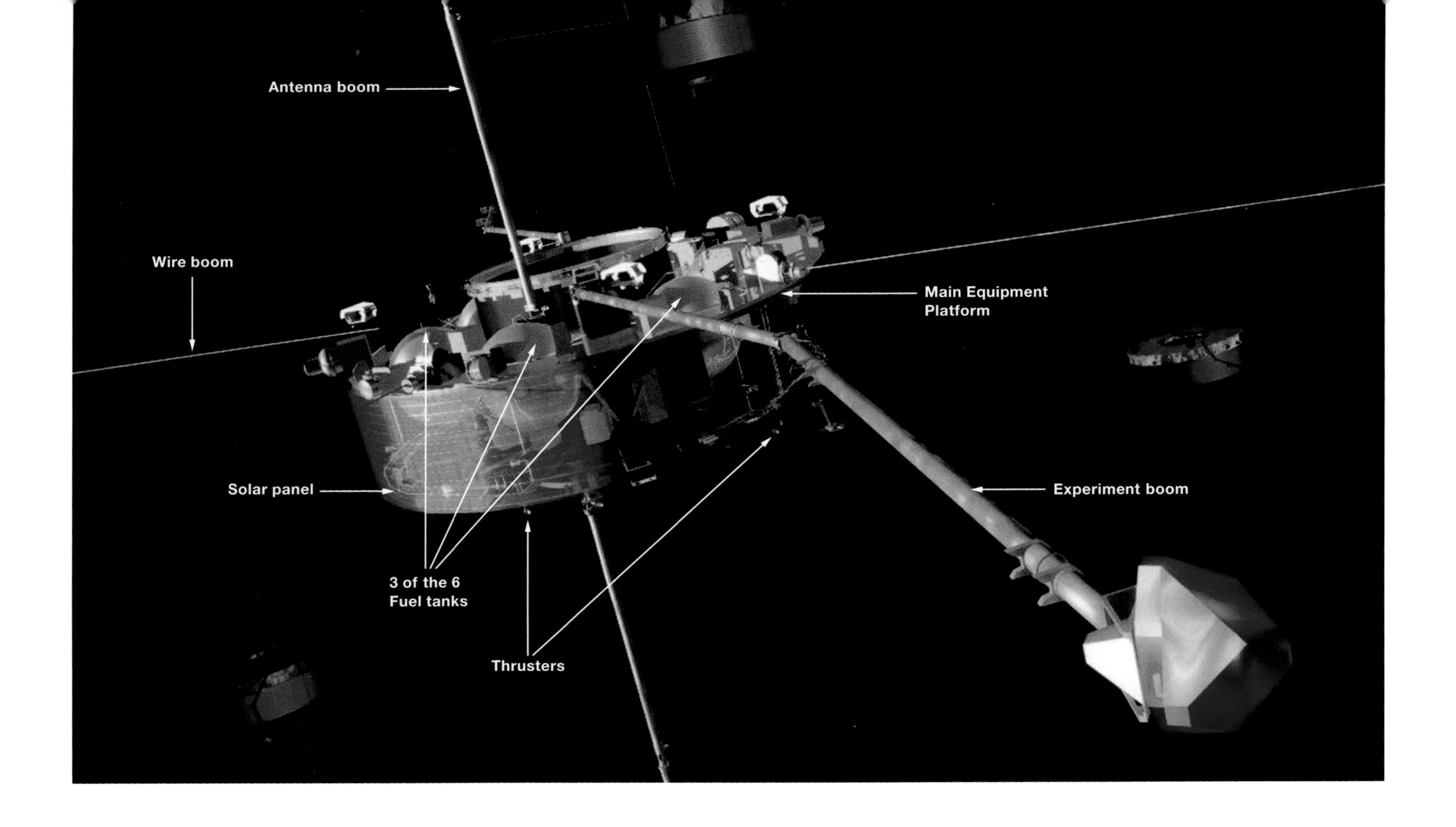

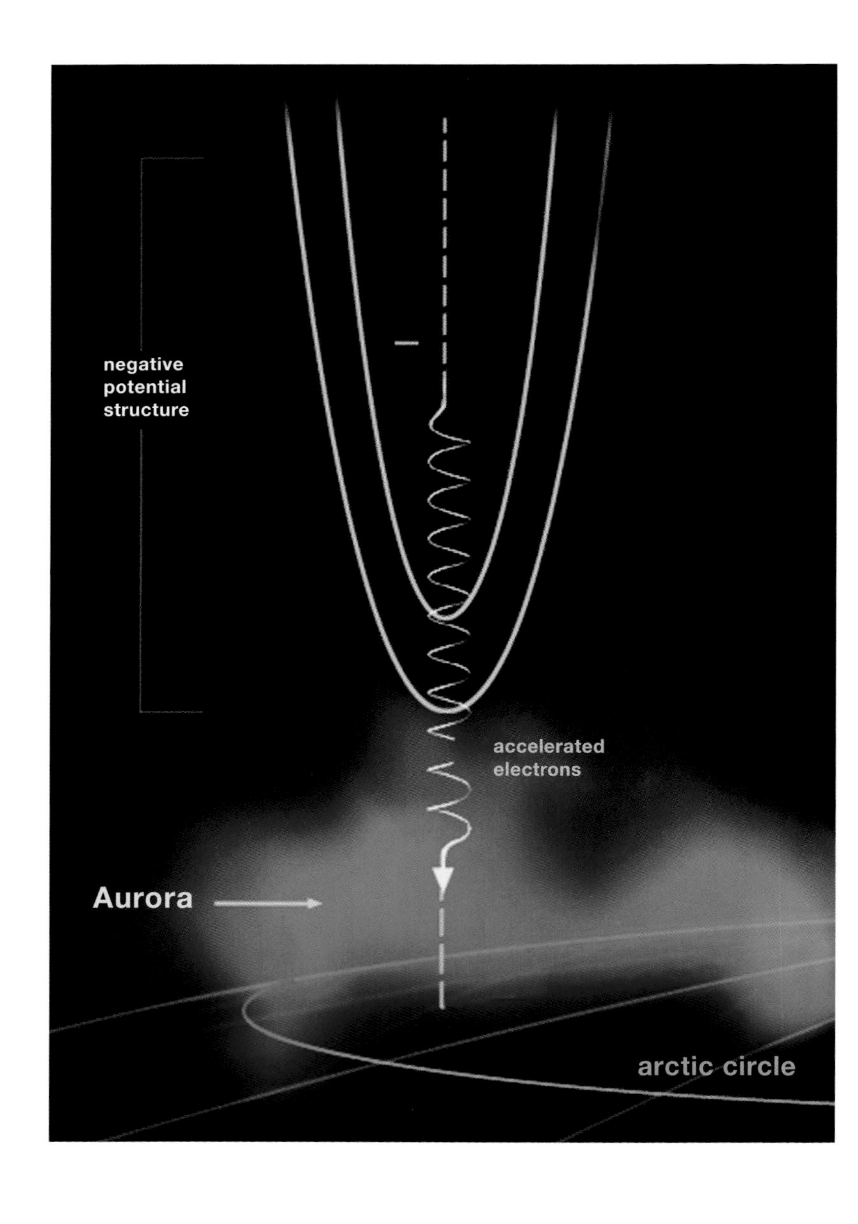

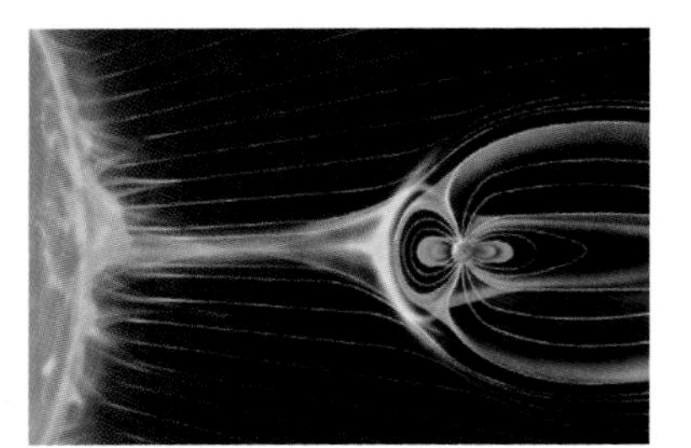

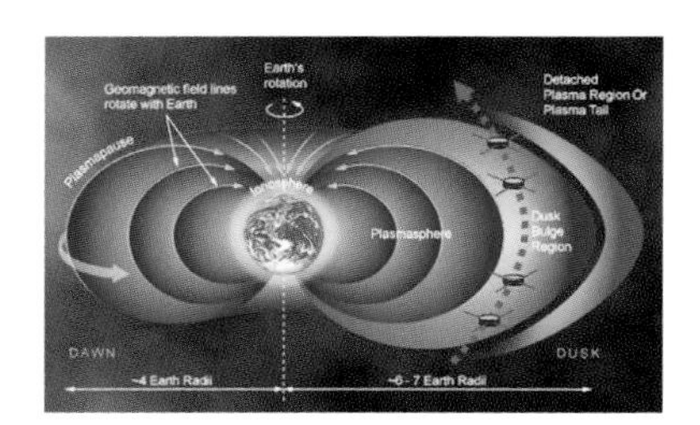

ABOVE
A cutaway drawing of one of the four identical Cluster spacecraft.

RIGHT
The four Cluster spacecraft help scientists understand the mechanism of the Earth's auroras, illustrated here as electrons spiraling down magnetic field lines until they encounter the upper atmosphere, producing light.

FAR RIGHT, TOP
An artist's rendering of the relationship between solar winds, solar flares, and the Earth's protective response.

FAR RIGHT, CENTER
Using graphic elements, an artist shows the interface between the solar wind and the Earth's magnetosphere.

FAR RIGHT, BOTTOM
An illustration showing the Cluster spacecraft as they fly from south to north through the outer portion of the plasmasphere.

of Daimler-Chrysler Aerospace) geared up to produce the four replacement satellites in November 1997, and the Russian consortium Starsem agreed to carry the payload. On July 16, 2000, two of the Clusters entered orbit after being launched from the Baikonur Cosmodrome, Kazakhstan; the other two followed suit on August 9, 2000. Eighteen years after conception, ESA got the system it wanted. Its initial operational lifespan of two-and-a-half years was surpassed easily.

The Cluster satellites conceived by the ESA scientists plumb the mysteries of the Earth's magnetosphere. Each of the small circular spacecraft measures just 9 ft 6 in. (2.9 m) in diameter and 4 ft 4 in. (1.3 m) in height, and weighs only 2646 lb (1200 kg), including 1433 lb (650 kg) of propellant necessary for frequent maneuvers, and a mere 157 lb (71 kg) for scientific payload. The mission explores the physical relationship of the Sun to the Earth. Flying together, the Cluster quartet observes and measures minute changes in that part of space adjacent to our planet, looking for the influence of subatomic charged particles associated with solar wind as they collide with the world's magnetic coat of armor, the magnetosphere. Cluster's indistinguishable instruments enable simultaneous measurements; this has produced the first three-dimensional maps of the magnetosphere as a region responds to bombardment by the solar wind. Cluster also acts as an early warning system to Earth-bound activities sensitive to the Sun's influences, such as the generation of electrical power, communications, satellite control, and weather forecasting.

One of the two Cluster pairs is lowered from a vertical to a horizontal position prior to encapsulation at Baikonur, Kazakhstan, in August 2000.

Cluster is a truly international project, involving more than two hundred co-investigators from such countries as the European Union member states, the U.S., China, Canada, India, Israel, and Russia. Each satellite carries a suite of eleven instruments:

- Fluxgate Magnetometer (FGM), Imperial College, U.K.: measures the magnetic field during the orbits, at a rate of up to 67 samples per second.
- Active Spacecraft Potential Control (ASPOC) experiment, Space Research Institute, Austria: neutralizes the intense exposure of Cluster's scientific instruments by preventing increases in positive electrical charges.
- Electron-Drift Instrument (EDI), Max Planck Institute, Germany: determines the electrical and magnetic field strength near the spacecraft.
- Spatio-Temporal Analysis of Field Fluctuation (STAFF) experiment, Centre d'Étude des Environnements Terrestre et Planétaires, France: detects waves in the magnetic field.
- Electric Field and Wave (EFW) experiment, Swedish Institute of Space Physics: studies plasma convection and waves by measuring the Earth's electric field.
- Digital Wave Processing (DWP) experiment, University of Sheffield, U.K.: provides the computing for all wave experiments.
- Waves of High Frequency and Sounder for Probing of Electron Density by Relaxation (WHISPER) experiment, Laboratoire de Physique et Chimie de l'Environnement, France: assesses the density of charged particles.
- Wide Band Data (WBD) Instrument, University of Iowa, U.S.: measures electric and magnetic fields in certain bands.
- Plasma Electron and Current Experiment (PEACE), Mullard Space Science Laboratory, U.K.: evaluates low- to medium-energy particles in space plasma.
- Cluster Ion Spectrometry (CIS) experiment, Centre d'Étude Spatiale des Rayonnements, France: analyzes the functions of ions in space plasma and solar wind.
- Research with Adaptive Particle Imaging Detectors (RAPID), Max Planck Institute, Germany: a particle detector that senses the highest energy ions and electrons entering it from space.

After two extensions to Cluster's initial two-and-a-half-year life, its mission is planned to end in December 2009.

The control room at the European Space Operations Center in Darmstadt, Germany, where the Clusters are monitored.

OPPOSITE
Encapsulated and ready for launch, two of the Cluster satellites sit atop their booster on the pad at Baikonur.

CLUSTER
STARSEM

Exploring the Solar System

Voyages of Discovery

3

Introduction

On December 3, 2006, one of the ESA's *Rosetta*'s instruments recorded this image of Mars (top) amid the Milky Way. Mars appears disproportionately large, owing to overexposure of the image.

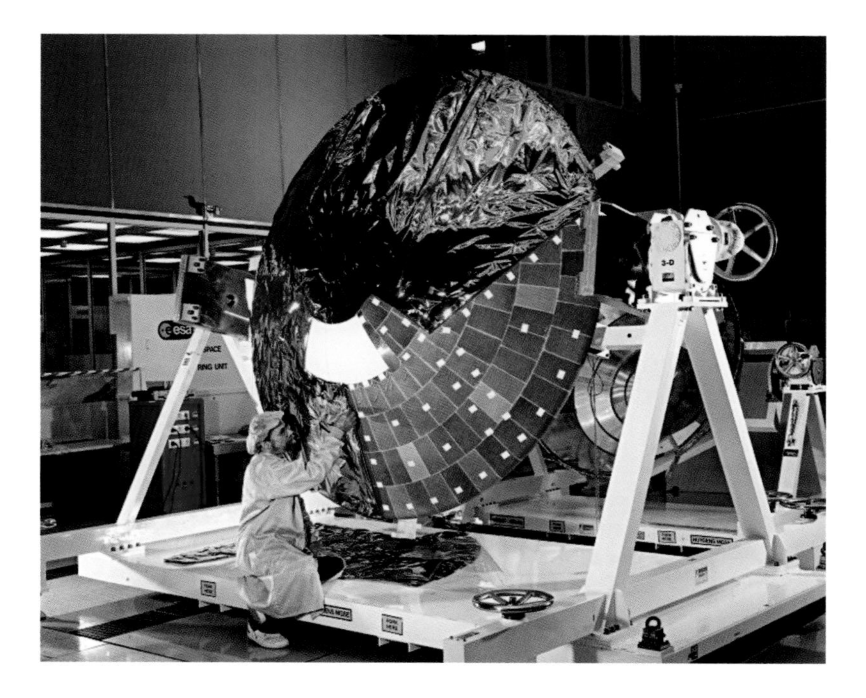

ABOVE
A technician applies sheathing to the heat shield of the European Space Agency's *Huygens* probe before its integration with the Jet Propulsion Laboratory's *Cassini* spacecraft.

BELOW, LEFT AND RIGHT
Two images of Mars captured by the *Rosetta* spacecraft, one in near infrared and near ultraviolet (left), the other produced by enhancing the ultraviolet signal.

Beginning with America's first successful satellite, *Explorer I*, launched in January 1958, the nation learned about and became intrigued with a Southern California institution known as the Jet Propulsion Laboratory (JPL). Engineers at JPL designed *Explorer I*, the U.S. answer to Sputnik. Owing to its resulting fame, the name of this obscure science facility conjured up in the public mind almost as much excitement as did astronauts and space suits.

JPL began during the late 1930s as an offshoot of another prodigy of the Golden State, the California Institute of Technology (Caltech). The concept of JPL began with Hungarian-American faculty member Theodore von Karman, founder of the Guggenheim Aeronautical Laboratory at Caltech. His lab hosted perhaps the first rocketry program on any U.S. campus. During the Second World War, the U.S. Army Ordnance Department funded JPL's rocket research, and continued to pay the bills of the growing complex in the Pasadena foothills.

With the founding of America's national space agency, NASA, in 1958, JPL won a new role as the agency's center for interplanetary spaceflight. JPL's engineers and scientists did not take long to fulfill this mission. The Mariner probes launched during the 1960s flew by Mars and captured the country's imagination with clear photographs of this distant world. The Ranger and Surveyor missions to the Moon also bore the indelible JPL stamp. Of course, these all reflected NASA achievements, but for space enthusiasts and casual observers, only JPL epitomized these thrilling adventures to the Solar System.

Pioneer 10 and 11

Pioneer 11 is given a checkout prior to launch.

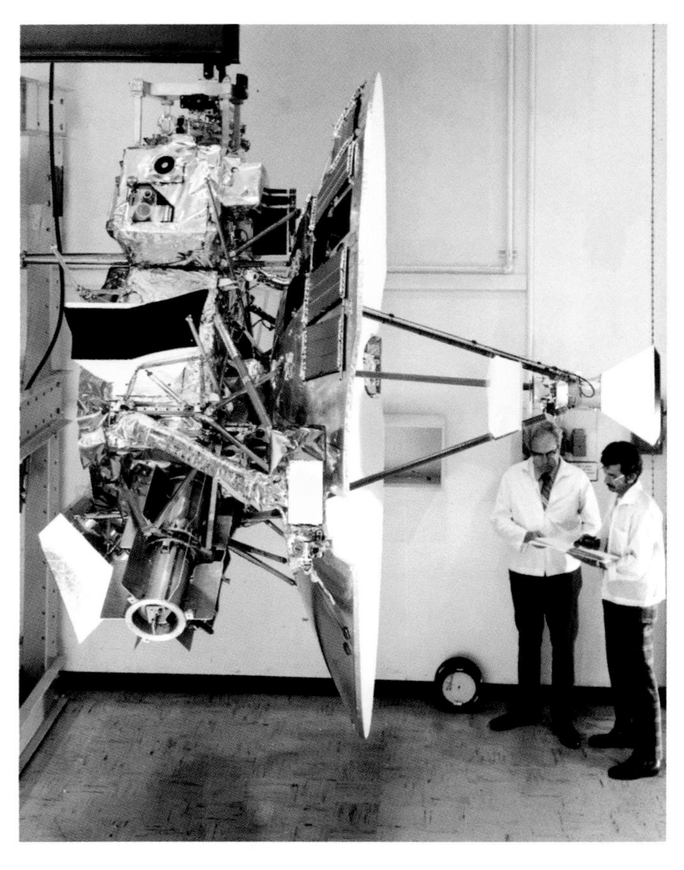

Technicians conduct pre-launch testing on the upper portion of *Pioneer 10*.

Key facts

MANAGEMENT:	**NASA Ames Research Center**
MANUFACTURER:	**Thompson Ramo Wooldridge Company**
LAUNCHED:	***Pioneer 10*, March 1972; *Pioneer 11*, April 1973**
END OF MISSION:	***Pioneer 10*, March 1997; *Pioneer 11*, September 1995**
PURPOSE:	**Fly by Jupiter, Saturn, and Neptune, after which collect data as they left the Solar System**
WEIGHT:	**570 lb / 259 kg each**
PROPULSION:	**Heat generated by nuclear materials**

Gradually, other NASA centers managed planetary missions as well, joining JPL at the forefront of such exploration, but certainly not eclipsing the original. Indeed, some thought that different centers ought to win these projects, if only to give JPL some competition. An early example of this trend occurred in the long line of Pioneer spacecraft. Beginning in late 1965, NASA Ames Research Center in Sunnyvale, California, managed *Pioneer 6*, and all of the subsequent Pioneers. *Pioneer 6* to *9* flew in interplanetary space along the Earth's solar orbit, with the responsibility of predicting solar storms. The small, lightweight *Pioneer 10* and *11* had a very different mission, the result of which has kept them in flight to the present day.

Many of NASA's scientists dreamed as early as the late 1960s of a flight to the giant Jupiter, with twice the mass of all the other planets combined. This adventure had at least one great problem, aside from Jupiter's distance from the Earth: its distance from the Sun. It was too far away to make solar panels of any use, so the manufacturer of *Pioneer 10* and *11*—the Thompson Ramo Wooldridge Company—fabricated the world's first probes fueled by the heat generated by nuclear materials. These long-distance travelers measured only 9 ft 6 in. (2.9 m) in length, had a 9-ft (2.7-m) high-gain antenna, and weighed just 570 lb (259 kg) apiece.

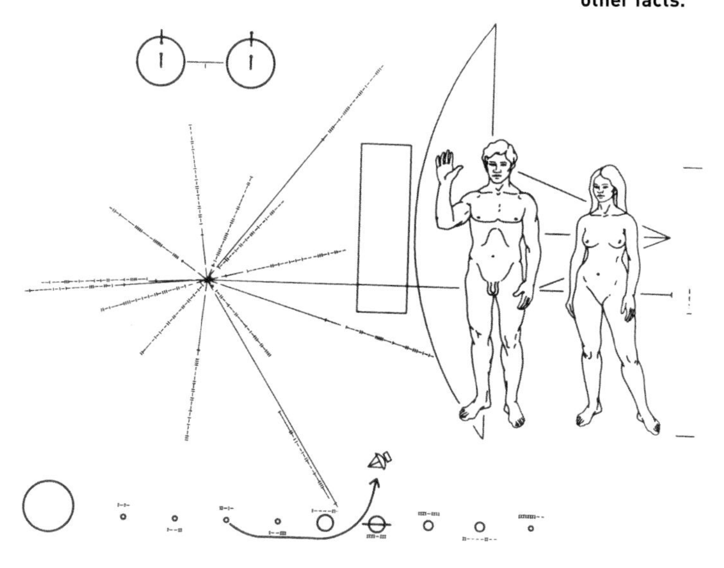

Attached to the exterior of each of the Pioneer spacecraft, this pictorial plaque gives whomever the probes might encounter representations of human anatomy and stature, the Solar System, and many other facts.

This photograph of Jupiter was taken by *Pioneer 11*, which reached the planet before *Pioneer 10* and flew much closer than its counterpart.

Art imitates life: artist Wilson Hurley's oil painting of Saturn, featuring the *Pioneer 11* spacecraft.

***Pioneer 11* recorded this image of Saturn during its flyby. The spacecraft also took measurements of the ringed planet's atmosphere.**

As *Pioneer 10* left the Earth's grip on March 3, 1972, an Atlas–Centaur rocket combination boosted the spacecraft to the fastest speed (32,188 mph/51,796 kmph) of any object launched up to that time. *Pioneer 10* soon accomplished another milestone, becoming the first probe to venture beyond Mars. As it went between the orbits of Mars and Jupiter, the project team got lucky: although the spacecraft suffered some hits, it got through the asteroid belt between these planets intact. *Pioneer 10* took pictures as it flew by Jupiter at its closest point in December 1973, crossed Saturn's orbit in February 1976, and continued sending home observations about the solar wind, the fringes of the Solar System, and the presence of cosmic rays from other parts of the Milky Way. NASA cut off regular contact with *Pioneer 10* in March 1997. The space agency continued to track *Pioneer 10* until its signal weakened to such an extent that it slipped from detection in February 2003. Engineers made an unsuccessful attempt to pick up its signal in March 2006. Designed to survive for twenty-one months, this small spacecraft survived as a useful probe for thirty years, and continues to travel, carrying a greeting etched on an aluminum plaque that illustrates a man and a woman, the location of the Earth, and the date on which it left its home planet.

Pioneer 11 followed its sister ship into space in April 1973, flew first to Jupiter (safely across the Martian-Jovian asteroid belt), and came three times closer to the behemoth planet than *Pioneer 10* had done (26,600 miles/42,809 km). This pass, like *Pioneer 10*'s, enabled much better measurements of the severe Jovian radiation than any taken before. *Pioneer 11* reached Saturn in September 1979 and took pictures of the ringed planet from about 13,000 miles (20,921 km), discovered its surface to be made of liquid hydrogen, and found surface temperatures to be, on average, –290°F (–179°C). Then the probe began to travel in the opposite direction to *Pioneer 10*, and in 1990 sailed off beyond the edge of the Solar System, carrying the same plaque as that borne by its sister craft. Data about solar wind, the solar magnetic field, and cosmic rays gradually fell to a trickle. NASA retired *Pioneer 11* in September 1995, after its power source fell below the levels needed to continue scientific observations. Its trajectory is taking it toward the center of the Milky Way, and presumably far beyond.

Voyager 1 and 2

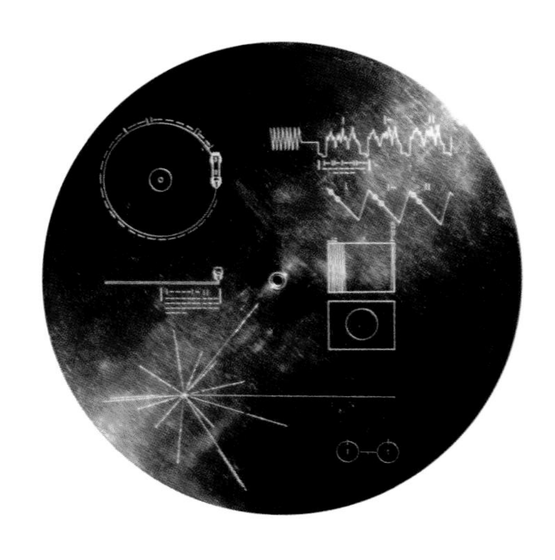

ABOVE
An artist's impression of one of the two identical Voyager spacecraft, still flying after more than thirty years of travel.

LEFT
The gold-plated phonograph record installed on the side of *Voyager 1* and *2*, containing speech, music, pictures, a variety of natural sounds, and much else from Earth.

RIGHT
One of the Voyager spacecraft, viewed from another perspective.

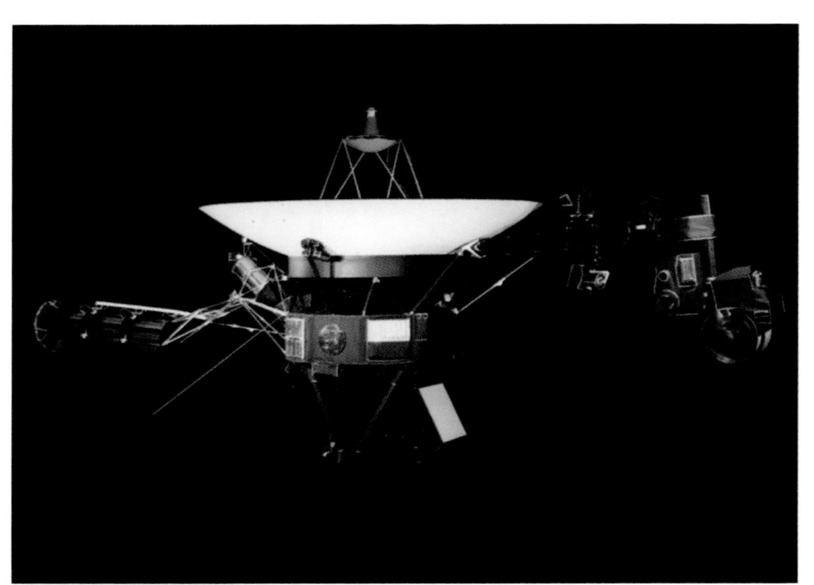

A technician standing next to one of the Voyager spacecraft cradles one of the two gold-plated records, one for each spacecraft, on which researchers had etched many of Earth's characteristic sounds and sights.

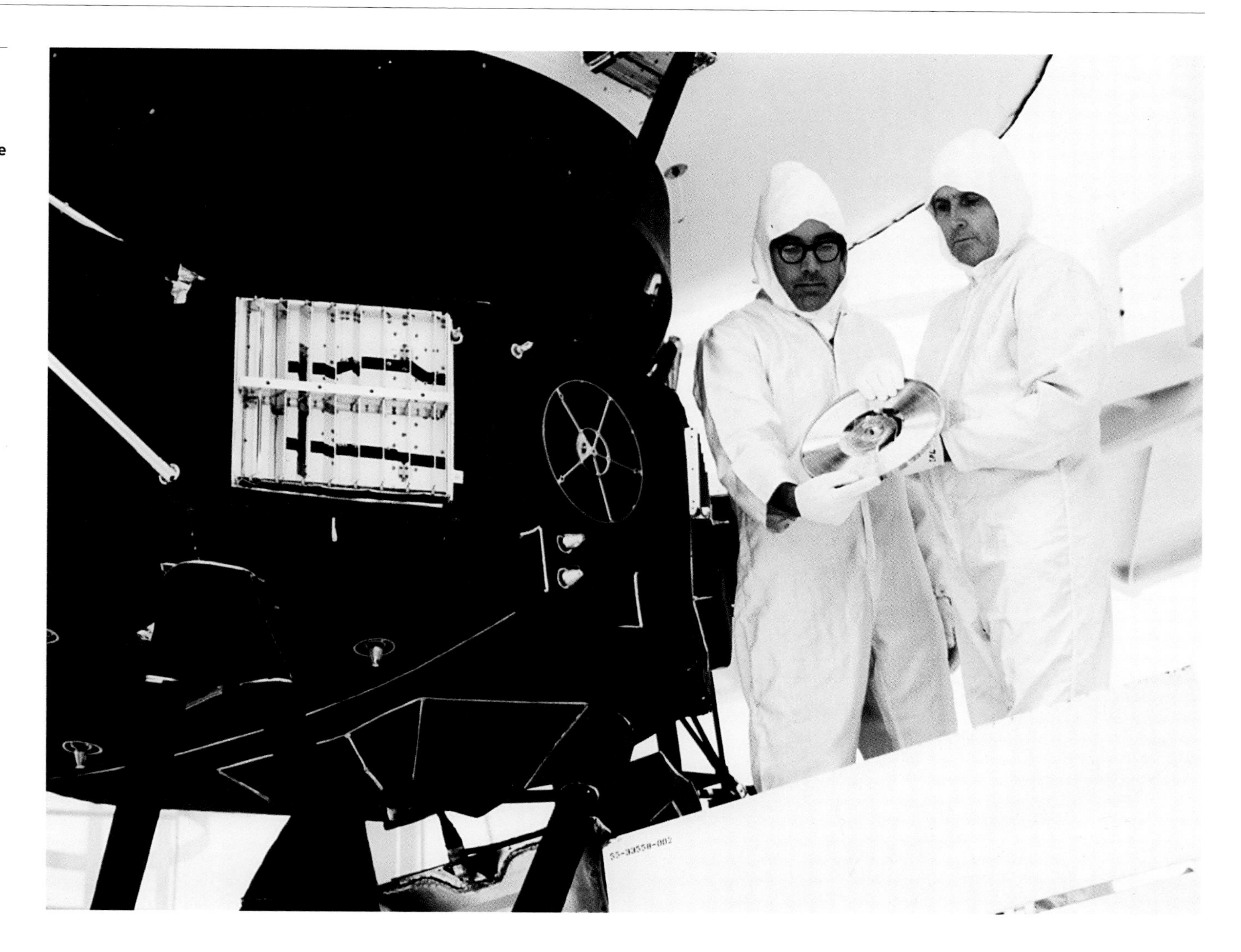

Key facts

MANAGEMENT AND MANUFACTURE:	**NASA Jet Propulsion Laboratory**
LAUNCHED:	***Voyager 2*, August 1977; *Voyager 1*, September 1977**
PURPOSE:	**Fly by and observe Jupiter, Saturn, Uranus, Neptune, their moons, and finally, penetrate interstellar space**
VOYAGER INTERSTELLAR MISSION:	**1989**
ESTIMATED LIFESPAN:	**Until 2020**
ESTIMATED DISTANCE FROM EARTH AT END OF LIFESPAN (VOYAGER 1):	**12.4 billion miles / 20 billion km**

JPL burnished its reputation in planetary travel with a bold initiative called the Grand Tour, hastened by the alignment of the outer planets between 1976 and 1979—an opportunity that only presents itself every 175 years. It enabled the probes to swing past the planets by gravity-assist (picking up speed as they flew by), thus reducing the travel time and eliminating the need for heavy and elaborate onboard propulsion, and to reach more distant bodies in the Solar System. Indeed, mission planners plotted the Grand Tour so that the spacecraft would fly past not only all of the Solar System's big planets (Jupiter, Saturn, Uranus, and Neptune), but also forty-eight of their moons. As if to underscore JPL's leadership, its engineers decided to rely on a proven design (an advanced version of the Mariner), but to conceive and fabricate the Grand Tour probes right on the Pasadena campus, rather than contract out to industry.

Known as *Voyager 1* and *2*, these identical spacecraft arrived on schedule at Kennedy Space Center in summer 1977. *Voyager 2* flew first (August 1977), aboard a Titan III–Centaur rocket combination, on its sojourn to Jupiter, Saturn, Uranus, and Neptune. In April 1979, it transmitted superb photos of Jupiter and its moons (Amalthea, Io, Callisto, Europa, and Ganymede), in addition to motion pictures of the swirling Jovian atmosphere. *Voyager 2* next visited Saturn, where the planet's rings underwent photographic scrutiny, as did its moons Hyperion, Enceladus, Tethys, and Phoebe. By January 1986, the spacecraft passed Uranus where, incredibly, it discovered ten new moons. Finally, the first ever flyby of Neptune occurred in August 1989, when *Voyager 2* penetrated to within 2800 miles (4506 km) of its surface. It discovered five new moons, an atmosphere of hydrogen and methane,

Two spectacular photographs of Jupiter taken by *Voyager 1* and enhanced by the JPL team's use of filters and multiple images. The two Voyager spacecraft took a total of 33,000 images of Jupiter and photographed five of its moons. The pictures here highlight the massive planet's Great Red Spot.

Voyager 2 also approached far-off Neptune, seen here with high-altitude white clouds and its Great Dark Spot (right), as well as from a much greater distance.

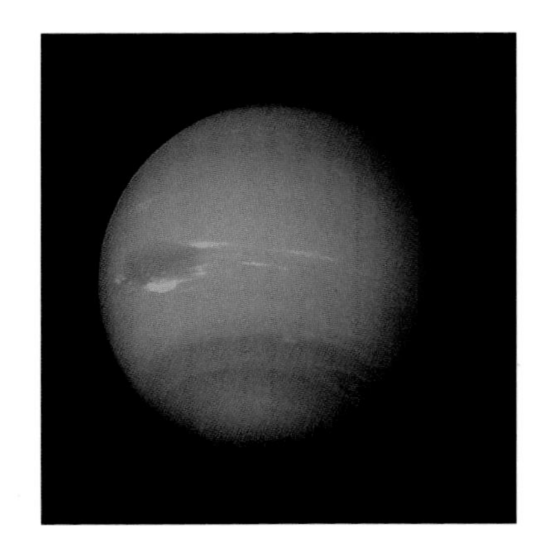

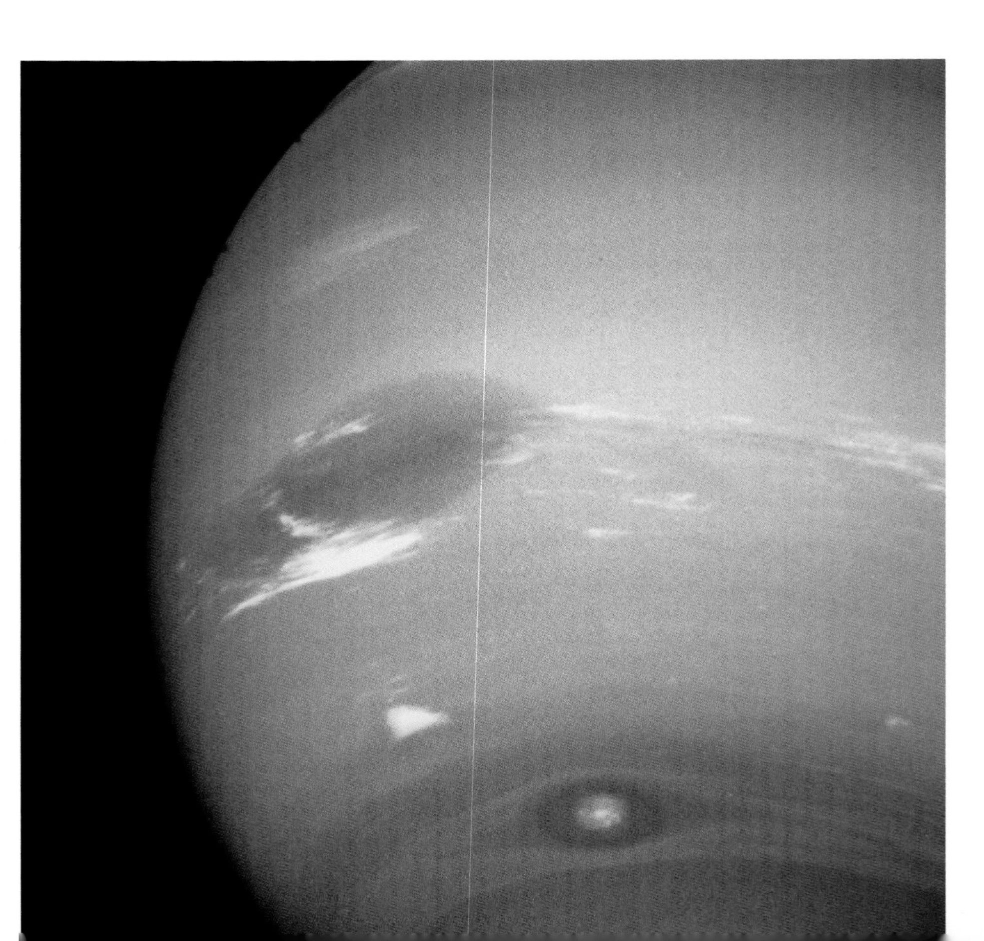

and winds as high as 680 mph (1094 kmph). *Voyager 1* lifted off in September 1977 (two weeks after its twin) and actually reached Jupiter and Saturn before its sister ship because it took a more direct pathway; it was *Voyager 1* that first alerted astronomers to a Jovian ring in 1979. Traveling close by Saturn's moon Titan, *Voyager 1* sped into open space.

As both spacecraft pushed out from the Solar System in 1989 (twelve years after being launched), they entered a phase of their journeys known by NASA as the Voyager Interstellar Mission (VIM). As they fly on, they are prepared for encounters with the unknown. Each one has on board a gold-plated, 12-inch phonograph record containing images, music, spoken languages, and such sounds as a horse and cart, a chimpanzee, Morse Code, laughter, and a heartbeat. Beyond such chance meetings, the VIM mission team hoped to acquire data about the region of space at the limits of the Sun's influence. The scientists wanted to define the boundary of the heliopause, the outer edge of the Sun's magnetic field, where interstellar space begins, characterized in part by the slowing of the solar wind from supersonic to subsonic speed and the expansion of the interstellar wind. Barring unforeseen events, both spacecraft should continue to function at least until the year 2020, by which time their nuclear battery power pack and thruster fuel may be expended. By then, *Voyager 1* will have traveled 12.4 billion miles (20 billion km) from the Earth, and *Voyager 2* will be 10.5 billion miles (16.9 billion km) distant. In time, both will wander the Milky Way galaxy. In the meantime, they have become the most distant objects fashioned by human beings, a hundred times farther from the Sun than is the Earth. At this phase of their journey—at the very rim of the Solar System, in a region called the heliosheath—a backward glance toward the Sun by either Voyager would show nothing more than a bright point of light.

A remarkably clear picture of the stately Saturn, taken by a Voyager spacecraft during its long journey.

Mars Global Surveyor

Pioneer and Voyager represent exceptions to planetary exploration up to the early twenty-first century. For the most part, scientists and engineers launch spacecraft that either orbit around or land on bodies in the Solar System. Flybys that culminate in long-distance travel happen rarely. But regardless of the means of data collection and observation in this process of discovery, no object in the skies has received greater interest or been the subject of greater enthusiasm than Mars, the so-called Red Planet. Because of such factors as similarity to the Earth in size (half that of our planet), position relative to the Sun (fourth, in contrast to the Earth's third), and the composition of its atmosphere (carbon dioxide), Mars has been the subject of intense speculation for hundreds of years.

In 1996, NASA mounted its first successful mission to Mars in two decades, launching *Mars Global Surveyor* on November 7 of that year. (*Viking 1* and *2* orbited and landed in 1976.) *Surveyor* left the Earth's atmosphere atop a Delta rocket, the first time the relatively economical launch vehicle had powered a flight to another planet. It entered the orbit of Mars on September 12, 1997, and for the next year-and-a-half *Surveyor*'s mission controllers slowly adjusted its broad elliptical orbit into a low, nearly circular one over the planet's poles. Beginning in March 1999, *Surveyor* pursued its mission, involving no less than mapping the Martian atmosphere, surface, and interior. Its first phase of exploration ended in late January 2001, after which it began an extended period of research.

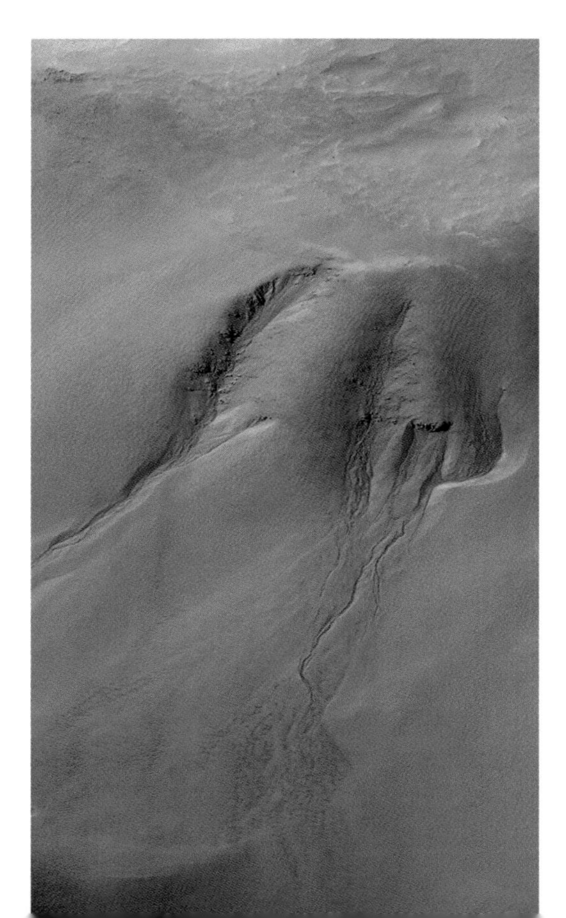

***Mars Global Surveyor*'s Orbiter Camera took this high-resolution image of gullies on the walls of a meteor crater. The associated markings suggest water action.**

Mars Global Surveyor—like most Martian expeditions—had four principal objectives:

1. Scientists wanted to clarify the debate about life on Mars. Did, or does, water, one of the essential preconditions of life, exist on the Red Planet, and did living creatures of some type propagate from it? Data from *Surveyor* suggests that liquid water might exist in some spots, just below the surface. It also detected evidence that water has flowed there in the past.
2. Researchers sought to characterize the Martian climate, which appears to be dominated by massive dust storms and seasonal changes in its ice caps, comprising frozen carbon dioxide and water. Mars' existing seasons and atmospheric interactions offer clues about the planet's historic climate, to which *Surveyor* contributed by assessing the planet over a full Martian year (687 Earth days).
3. Geologists hoped to grasp the physical attributes of Mars by analyzing such features as volcanoes, craters, rocks, and tectonic movement in order to lend insight into its origins. *Surveyor* discovered occasional magnetic materials as it flew overhead, implying that Mars once had a magnetic field.
4. Mission planners regarded *Surveyor* as preparatory to human trips to Mars. The search for water might result in finding a supply adequate for visitors, and a full comprehension of the Martian atmosphere might protect space travelers from such perils as intense radiation.

Fabricated for JPL by Lockheed Martin Space Systems, *Mars Global Surveyor* weighs 2342 lb (1062 kg), and measures about 10 ft (3 m) in length—just 4 x 4 ft (1.2 x 1.2 m) at its main body, which houses computers, radio systems, data recorders, and fuel tanks—with 40-ft (12.2-m) solar arrays. It consists of four instruments (some of which were adapted from the failed *Mars Observer* mission of 1992):

- The Mars Orbiter Camera provides a continuous photographic record of Martian weather. It features a high-resolution, narrow-angle lens capable of making images of objects as small as 1 ft 8 in. (0.5 m) across, enhanced by adjustments in the rotation of the spacecraft to complement the ground speed.
- A Thermal Emission Spectrometer maps mineral deposits using infrared radiation. It has discovered hematite, commonly associated with standing water.
- A magnetometer measures the planet's magnetic field in order to comprehend interior forces. The Martian magnetic field does not emanate from the core, but from small portions of the crust.
- Radio Science uses the onboard telecommunications system to measure variations in gravity.

Key facts

MANAGEMENT:	NASA Jet Propulsion Laboratory
MANUFACTURER:	Lockheed Martin Space Systems
LAUNCHED:	November 7, 1996
PURPOSE:	Explore the Martian climate, the geological history of Mars, and conditions for life, and lay groundwork for later human exploration
INSTRUMENTS:	The Mars Orbiter Camera, Thermal Emission Spectrometer, magnetometer
END OF SERVICE:	November 2006

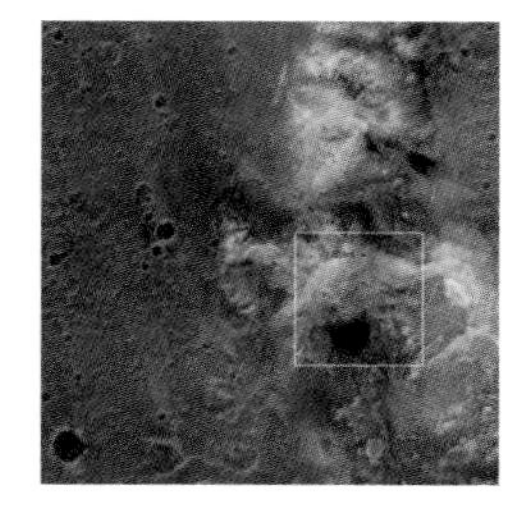

FAR LEFT
An artist's rendering of the *Mars Global Surveyor* in flight.

About two years after Mars Rover *Spirit* landed in Gusev Crater, *Mars Global Surveyor* found it in a field of view roughly 2 sq. miles (5.2 sq. km) in size.

On November 2, 2005, *Mars Global Surveyor* pinpointed the wandering *Spirit*.

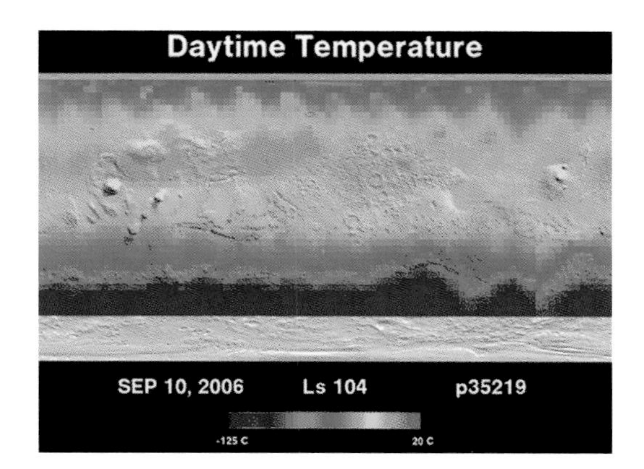

Acting like a weather satellite, *Mars Global Surveyor*'s Thermal Emission Spectrometer made daily weather maps of the Red Planet, such as this graphic of September 10, 2006.

In the end, *Surveyor* served its JPL team longer than any other spacecraft sent to Mars. Part of its longevity stemmed from a clever maneuver in August 2001, in which mission controllers inclined *Surveyor* backwards by 16 degrees relative to the surface, eliminating the need for thrusters to keep it on target, and resulting in a daily fuel saving of 800 percent. Passing year after year across Mars' surface, *Surveyor* was able to observe subtle changes in climate (shrinking carbon dioxide ice patches) and geology (movement of dunes) over time, thus providing a far more detailed and nuanced set of data than anyone had imagined in 1996. Finally, *Surveyor* quit. All seemed normal until November 2, 2006, when the spacecraft reported problems in one of its solar arrays. It went silent on the next orbit. By this time, it had flown four times longer than originally planned. A comprehensive report issued in April 2007 found death by battery failure, complicated by human miscalculations that disabled the solar arrays. Formal efforts to revive the spacecraft ceased on January 28, 2007, and *Mars Global Surveyor* remains in Martian orbit.

Mars Exploration Rover

A Rover is driven over ramps of uneven height to test the range of motion of the vehicle's suspension.

Early in the twenty-first century, a historic opportunity to see Mars in a way never before imagined presented itself. Scientists had realized that in August 2003, Mars and Earth would be closer to each other than they had been in thousands of years. Accordingly, in summer 2000, JPL leaders decided to send two revolutionary robots to Mars. Wisely, the designers incorporated features from *Sojourner*, a compact vehicle brought to Mars on July 4, 1997, during the Pathfinder project. *Sojourner* had roamed a small area of the planet successfully. The JPL team borrowed from *Sojourner* the six-wheel concept, a rocker-bogie suspension, a shell of airbags to absorb the landing, and a solar panel/battery power pack. But there the comparisons ended. These new Mars Rovers, as they came to be called, weighed 384 lb (174 kg), seventeen times more than *Sojourner*, and measured 5 ft 2 in. (1.6 m) in length and 4 ft 11 in. (1.5 m) in height: twice the dimensions of *Sojourner*. Not just conceived but also built at JPL, the first of the two robots left for its mission on a Delta II rocket fired from Cape Canaveral, Florida, on June 10, 2003. The second followed on July 7. Cocooned in its own aeroshell capsule, each one met the Martian atmosphere flying at 12,000 mph (19,310 kmph), then a parachute opened during the last two minutesof descent, followed by the inflation of the robot's airbags, a burst of retrorocket fire, and free fall to the surface. The first robot, called *Spirit*, following a national competition involving 10,000 students, landed first, on January 4, 2004, bouncing twenty-eight times before coming to rest. *Opportunity*, named in the same contest, dropped in on January 25 and stopped after twenty-six bounces.

The missions of *Spirit* and *Opportunity* had been carefully defined and separated. *Spirit* landed in Gusev Crater, a depression the size of Connecticut that may have been a lake at one time. *Opportunity* touched down on the other side of the planet, in Meridiani Planum, a flat expanse with deposits of hematite (a mineral formed in association with liquid water) on a landscape as big as Oklahoma. Each robot disentangled itself successfully from its landing site and rolled off to accomplish its tasks. To do so, they arrived equipped with five identical instruments:

- A panoramic camera, mounted on a 5-ft (1.5-m) mast with two lenses 12 in. (30.5 cm) apart, gives high-resolution, full-color, stereographic images. It takes landscape pictures of the Martian terrain, helping to pick targets as the vehicle progresses.
- A microscopic imager gives close-ups of such features as soils and rocks in order to determine their process of formation.
- A Miniature Thermal Emission Spectrometer (Mini-TES) scans the landscape and determines mineral deposits based on infrared wavelengths.
- A Moessbauer Spectrometer, located on the Rover's arm, measures the iron concentrations in certain minerals, and enables researchers to surmise what role water may have played in their formation.
- An Alpha-Particle X-Ray Spectrometer (APXS) detects—using traces of curium-244—major elements in the rocks and soils, which is essential in determining the origins of the samples.

Past and present. On the left is a model of the *Sojourner* spacecraft that preceded the Mars Rovers by about six-and-a-half years, and provided much design inspiration, as the photo suggests.

PAGES 112–13
A portrait of a Mars Rover, superimposed on the Martian terrain.

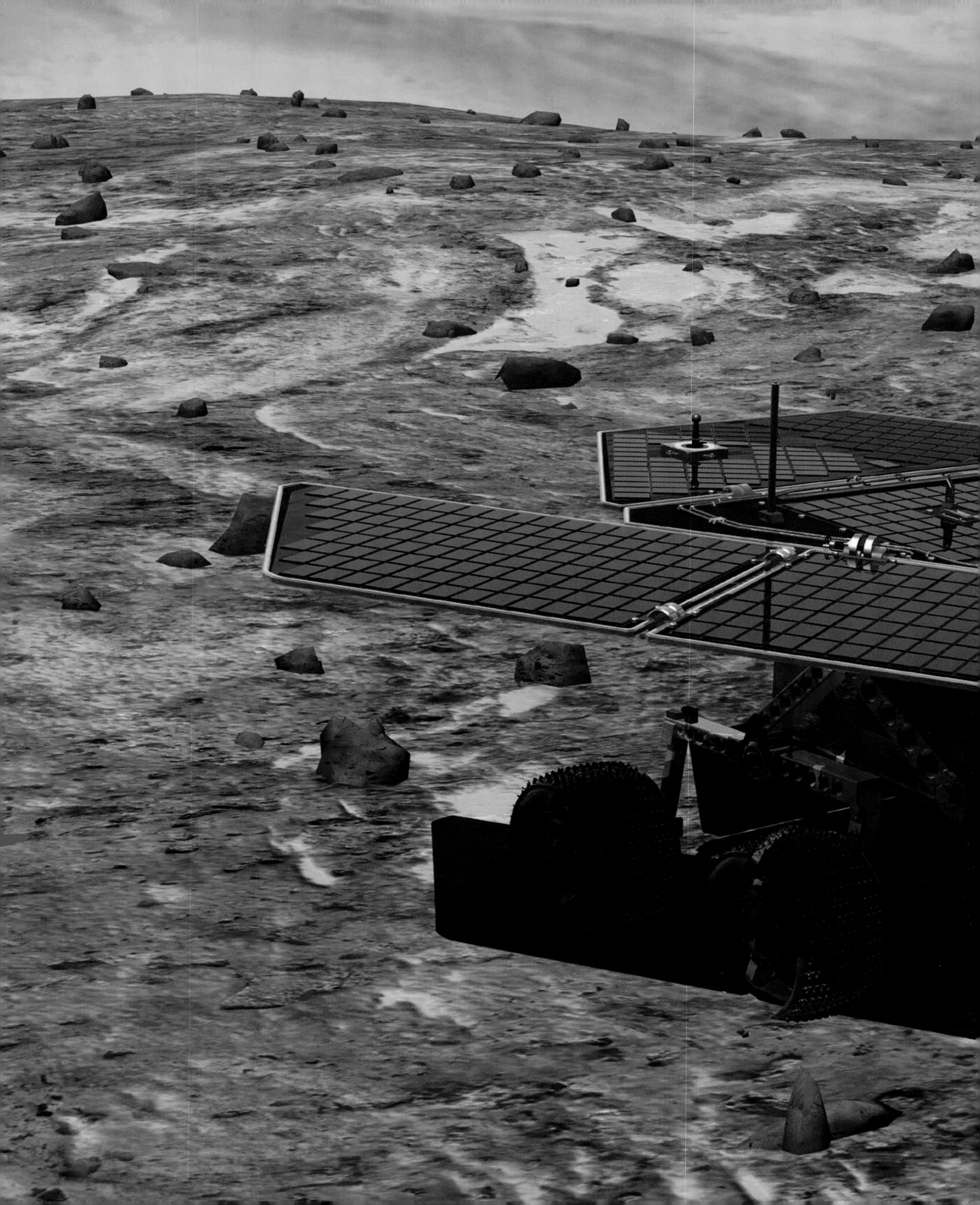

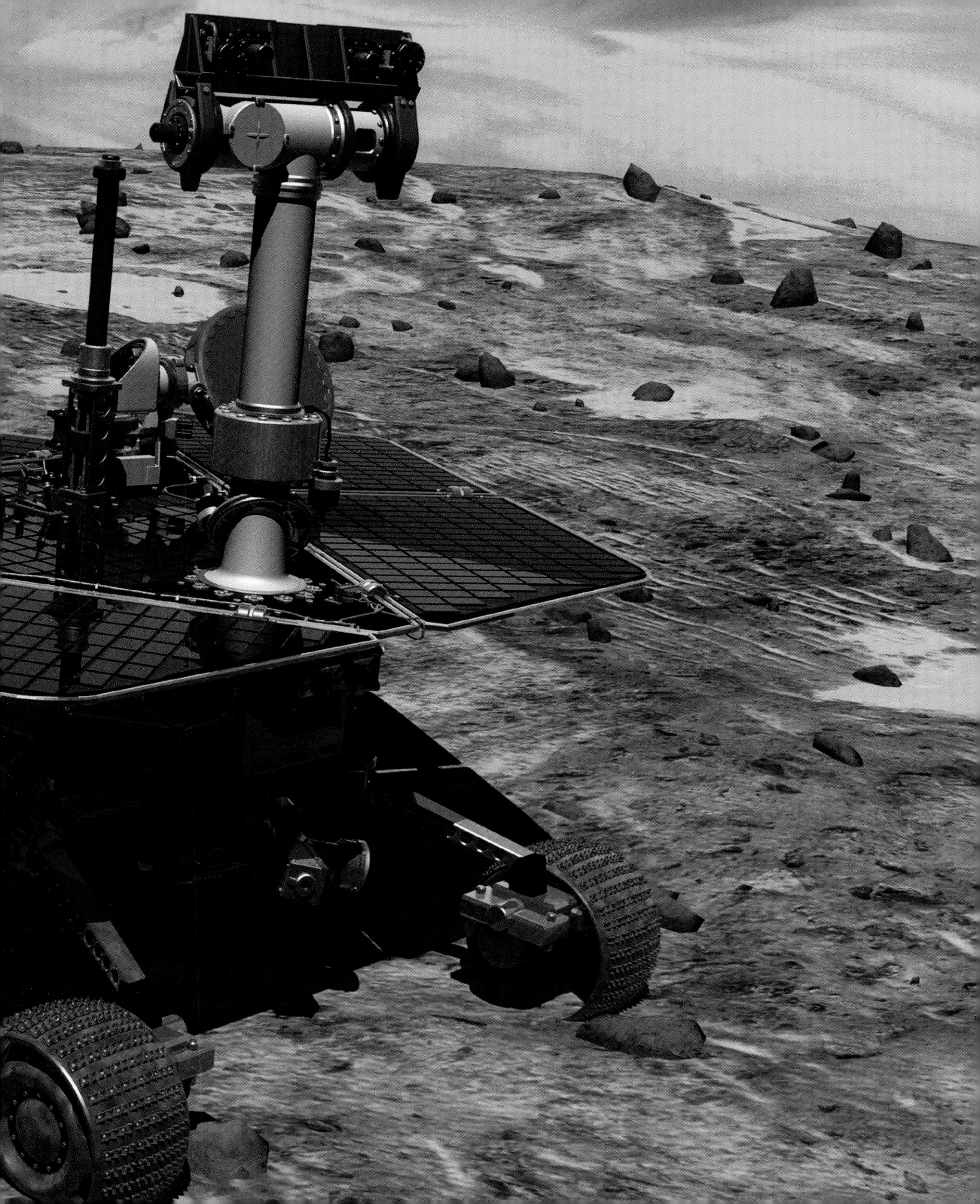

Key facts

LAUNCHED:	***Spirit* in June 2003, *Opportunity* in July 2003**
LANDING:	**Both in January 2004**
LANDING SITES:	***Spirit*: Gusev Crater; *Opportunity*: Meridiani Planum**
MANAGEMENT AND MANUFACTURE:	**NASA Jet Propulsion Laboratory**
INSTRUMENTS:	**Panoramic camera, Microscopic imager, Mini-TES, Moessbauer Spectrometer, APXS**
PURPOSE:	**Make geological surveys of opposite sides of the planet**

As the researchers released the Rovers on their sojourns across the Red Planet, they saw a group of hills from *Spirit*'s vantage point. These they (unofficially) named the Columbia Hills, after the ill-fated Shuttle *Columbia* mission, calling the seven hilltops Anderson, Brown, Chawla, Clark, Husband, McCool, and Ramon. Then the Rovers set off. Despite many hurdles (for example, repetitive re-booting of the computer on *Spirit*, and *Opportunity* becoming stuck in the sand for a month), the Rovers rolled past their expected three-month missions, enduring three-and-a-half years and counting. Although suffering from creaky joints owing to the extremes of the Martian climate, a frozen right front wheel on *Spirit*, and reductions in the range of motion in some of their moving parts, they continue to serve. *Spirit* moved 20 percent farther in one day (405 ft/123.4 m) than *Sojourner* did during its entire three months of travel on Mars. In seven months, *Spirit* navigated up to and over the Columbian Hills, on to a plateau of bedrock named Home Plate, into a ridge, and back to Home Plate. By spring 2007, it had traversed almost 4 ½ miles (7.2 km). Its main discovery has been evidence that water once changed the mineral composition of some soils and rocks on the landscape. *Opportunity* has toured several craters, including Eagle, Endurance, and the massive Victoria Crater, covering 6 ½ miles (10.5 km) in all. Its main finding has been that in the distant past, water saturated and flowed over part of the planet, according to mineral and rock texture evidence.

BELOW
The Mars Rover *Opportunity* recorded this vista as it stood at the Cape Verde promontory of Victoria Crater.

BOTTOM
An image of the Mars Rover *Opportunity* superimposed on the rim of Victoria Crater to convey the scale of the landscape.

ABOVE

This landscape is a composite of many images taken by the Mars Rover *Spirit* near its winter quarters.

BELOW

A top-down self-portrait of *Spirit* showing the deck of the spacecraft as well as the surrounding terrain.

ABOVE

At about 6.20 in the evening, Martian time, on April 23, 2005, *Spirit*'s panoramic camera made this image of Martian twilight at Gusev Crater.

BELOW

The panoramic camera on the Mars Rover *Spirit* took hundreds of images, combined by the JPL team into this single, 360-degree view from Husband Hill.

MESSENGER

As the twin Rovers drive through the Martian landscape, another NASA project seeks to learn more about Mercury, perhaps the least known among the planets. Its first encounters with a spacecraft occurred in March and September 1974, and again in March 1975, during three *Mariner 10* flybys, which captured pictures of the innermost planet to the Sun and mapped 45 percent of its surface at scales down to 3168 ft (966 m). About thirty years later (August 3, 2004), the *MESSENGER* spacecraft (an acronym for Mercury Surface, Space Environment, Geochemistry, and Ranging) went aloft from Cape Canaveral Air Force Station, Florida, atop a Delta II launch vehicle. In this combined mission sponsored by NASA, the Carnegie Institution of Washington, and the Johns Hopkins University Applied Physics Laboratory, the researchers wanted to learn more about the four, dense, so-called terrestrial (or rocky) planets containing iron cores: Mercury, Venus, Earth, and Mars. Regarding Mercury, the researchers had a number of questions: Why is it so dense? What is its geologic history? How is its magnetic core structured? How does its magnetic field function? What is the composition of the materials at its poles? What is its exosphere composed of?

MESSENGER's route to Mercury has hardly been straightforward. To avoid the cost of the heavy booster normally required for a direct flight (not allowed by NASA in its low-cost Discovery Program, of which *MESSENGER* is a part, anyway), scientists chose a seven-year voyage of 4.9 billion miles (7.9 billion km). This method employs gravity-assist to diminish or increase speed, rather than relying on mechanical acceleration or braking, which consumes a great deal of fuel and adds weight to the vehicle. As part of its solar orbit, in August 2005 the spacecraft traveled around the Earth for an initial gravity boost; in October 2006 and June 2007, it had flybys of Venus for another boost and a course correction, in preparation for three flybys of Mercury in January and October 2008 and September 2009. On these circumnavigations *MESSENGER* will take color images of most of the planet, including most of those parts missed by *Mariner 10*, and collect data relating to its atmosphere, surface, and magnetosphere. Finally, *MESSENGER* will enter the Mercurian orbit in March 2011. By then, the spacecraft will have completed fifteen orbits of the Sun.

Conceived and fabricated by the Johns Hopkins University Applied Physics Laboratory, *MESSENGER* is small in size (56 in./142 cm tall, 73 in./185 cm wide, and 50 in./127 cm deep) and light in weight (1101 lb/499 kg without propellant), which allows it to nestle behind an 8 x 6-ft (2.4 x 1.8-m) sunshade, necessary because of Mercury's proximity to the Sun. It features seven instruments:

- Mercury Dual Imaging System, to take pictures;
- Gamma-Ray and Neutron Spectrometer, to measure surface materials;
- X-Ray Spectrometer, to locate components of Mercury's crust;
- Magnetometer, to measure the magnetic field;
- Mercury Laser Altimeter, to measure surface features and determine whether the planet's core is solid or liquid;
- Energy Particle and Plasma Spectrometer, to detect charged particles in the magnetosphere;
- Mercury Atmospheric and Surface Composition Spectrometer, to analyze infrared to ultraviolet light.

OPPOSITE
An artist's conception of the encounter between the *MESSENGER* spacecraft and the planet Mercury when they meet in 2011.

A crane hoists the *MESSENGER* spacecraft out of NASA Goddard's Thermal Vacuum Chamber after tests subjected it to the extreme temperatures it will encounter going to, and orbiting, Sun-baked Mercury.

Key facts

LAUNCHED:	**August 2004**
MANAGEMENT:	**NASA, Carnegie Institution of Washington, Johns Hopkins University**
PURPOSE:	**Probe Mercury's geologic past, its high density, its magnetic core, its magnetic field**
ESTIMATED ARRIVAL IN MERCURY'S ATMOSPHERE:	**March 2011**
INSTRUMENTS:	**Mercury Dual Imaging System, Gamma Ray and Neutron Spectrometer, X-Ray Spectrometer, Magnetometer Laser Altimeter, Energy Particle and Plasma Spectrometer, Mercury Atmospheric and Surface Composition Spectrometer**

Deep Impact

BELOW
This artist's rendition depicts the Impactor being released by the flyby spacecraft, twenty-four hours before the collision. From left to right are Comet Tempel 1, the Impactor, and the *Deep Impact* flyby spacecraft.

BOTTOM
Artist Pat Rawlings imagines the moment when the Impactor met Tempel 1.

A drawing of the Impactor spacecraft, designed to navigate toward the sunlit portion of Tempel 1, to facilitate viewing by the flyby spacecraft and Earth-bound telescopes.

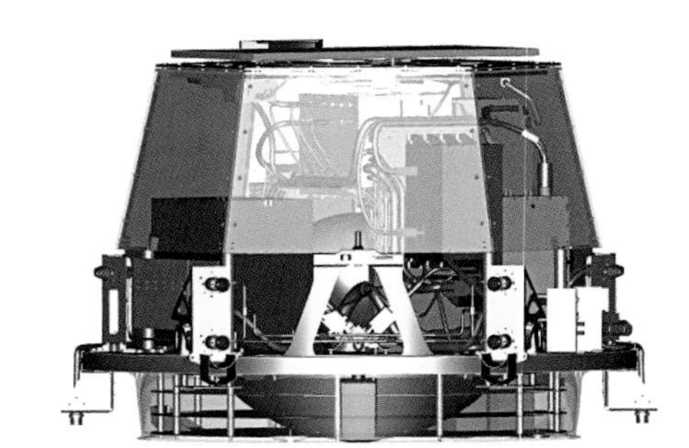

During the 1990s, as part of an orchestrated international attempt by researchers to comprehend the origins of the Solar System, a team of scientists conceived of a new way to uncover these secrets. Instead of digging at the surface like the Mars Exploration Rovers *Spirit* and *Opportunity* (see pp. 110-15), or flying nearby, this technique required that a vehicle pinpoint a target precisely on a celestial body and strike it with such force that it penetrates the body's surface. A demonstration of this technique was carried out using a vehicle named *Deep Impact*. NASA mission controllers trained *Deep Impact* on Comet Tempel 1, chosen because it could be reached easily and because it had a nucleus and mass large enough to withstand a crash without disintegrating. It measures about 45 sq. miles (117 sq. km) in size.

Deep Impact represented an alliance between JPL and the University of Maryland. Planning for the mission took place from 1999 to 2001, after which period the fabrication of the vehicle and instruments began. Actually, *Deep Impact* consisted of two vehicles mated together: a bigger, flyby spacecraft designed to reach Tempel 1, and a smaller vehicle called an Impactor. The flyby spacecraft, the size of an average Sports Utility Vehicle (10 ft 10 in./3.3 m long, 5 ft 7 in./1.7 m wide, 7 ft 6 in./2.3 m high, and 1325 lb/601 kg at launch), flew through space, controlled by a group of thrusters fueled by hydrazine. It carried just two scientific instruments, both designed to witness and record the Impactor's violent encounter with Tempel 1. Its large High Resolution Instrument consisted of a telescope with an 11¾-in. (30-cm) diameter that beamed light to an infrared spectrometer, as well as to a multispectral camera (perhaps the most powerful yet to fly in deep space), focusing on the site of the collision. The smaller Medium Resolution Instrument—a telescope just 4¾ in. (12 cm) in diameter—covered wide-angle shots to view the field of material ejected by the encounter.

The Impactor, just 39 x 39 in. (99 x 99 cm), weighed 820 lb (372 kg) at launch. Of that mass, 249 lb (113 kg) consisted of copper plates machined into a spherical shape and mounted at the collision end of the little spacecraft, providing the necessary weight to give it maximum force for burrowing into Tempel 1. Powered by only a 250-amp battery, it shared a good many of its sister ship's subsystems. It had just one instrument, a telescope like the smaller one on the flyby spacecraft, which provided images as the comet came closer. The great *Spitzer Space Telescope* (see pp. 172–75) also trained its eye on the event.

Key facts

MANAGEMENT:	**NASA Jet Propulsion Laboratory and the University of Maryland**
LAUNCHED:	**January 2005**
PURPOSE:	**Assess the composition of Comet Tempel 1**
COMPONENTS:	**Flyby spacecraft (1325 lb / 601 kg); Impactor spacecraft (820 lb / 372 kg)**
INSTRUMENTS:	**On the flyby vehicle: high- and medium-resolution telescopes; on the Impactor: medium-resolution telescope**
IMPACT:	**July 2005**

The twin explorers lifted off from Pad 17-B at Cape Canaveral Air Force Station, Florida, on the afternoon of January 12, 2005, aboard a Delta II launch vehicle. Nearly 83 million miles (134 million km) and a little more than five months later, the flyby ship began to take continuous images of Tempel 1. On July 2, 2005, mission controllers altered its trajectory and awakened the Impactor's battery. The next day, the Impactor flew on its own power and began to send images home. After three targeting maneuvers on July 4, the Impactor struck its target at 23,000 mph (37,000 kmph).

The resulting cataclysm created an immense Fourth of July spectacle, with a force equivalent to 9600 lb (4350 kg) of TNT, gouging a huge crater in Tempel 1 and casting a gigantic plume that reflected sunlight, rendering it visible from the Earth (though faintly) with the unaided eye. As the cloud of particles rose, far-off *Spitzer* and the nearby *Deep Impact* spacecraft peered into the crater, discerning some expected cometary ingredients, such as silicates, but also some unanticipated ones, such as clay, carbonates, iron-bearing compounds, and hydrocarbons. There have been other surprises, too, as scientists comb the data, including a wealth of organic matter inside the comet, 7 acres (2.8 ha) of dirty ice on its surface, and the possible siting of Tempel 1's birthplace—somewhere in the vicinity of Neptune and Uranus. Meanwhile, the *Deep Impact* flyby spacecraft continues to orbit the Sun, in sleep mode but capable of being awakened for further exploration.

Deep Impact's Impactor being assembled at Ball Aerospace and Technologies Corporation of Boulder, Colorado.

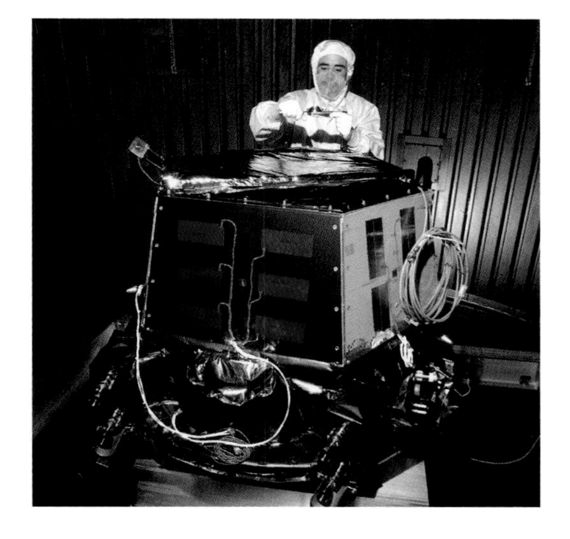

RIGHT
An engineer at Ball Aerospace and Technologies Corporation installs the telescope for the Medium Resolution Instrument that accompanied the *Deep Impact* flyby spacecraft.

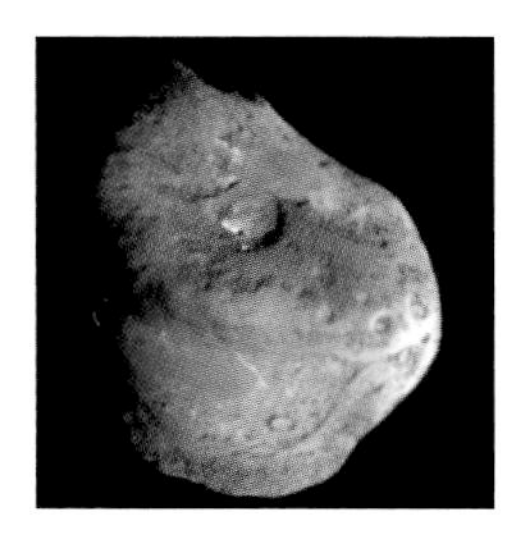

The *Deep Impact* Impactor captured this image of Comet Tempel 1 just five minutes before its collision with the comet.

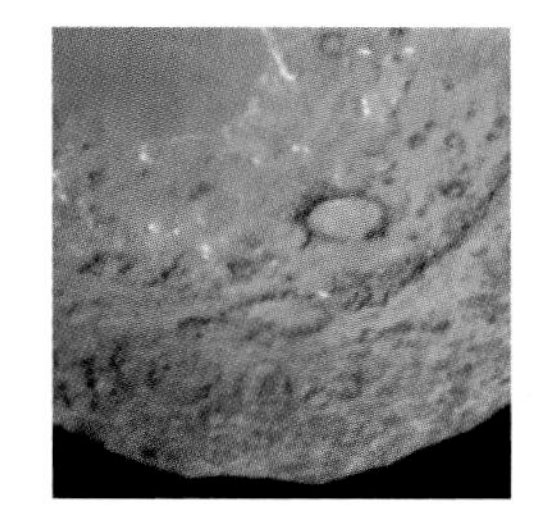

This picture of Tempel 1 was recorded by the Impactor just ninety seconds prior to their violent encounter.

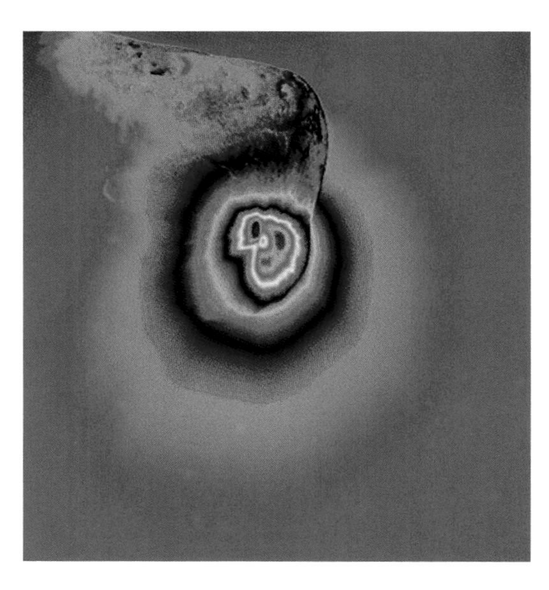

LEFT
The collective data of the impact is illustrated in this brightly colored graphic, indicating the nucleus (blue and maroon) and the impact flash (ringed in multiple colors).

Mars Reconnaissance Orbiter

An illustration of the *Mars Reconnaissance Orbiter* as it might look in orbit, with the large silver-colored high-gain antenna on top and the spacecraft's main bus pointed toward Mars.

Key facts

MANAGEMENT:	**NASA Jet Propulsion Laboratory**
LAUNCHED:	**August 2005**
MARTIAN ORBIT:	**March 2006**
PURPOSE:	**Make close observations of the Martian surface, accompanied by very high-volume data collection**
INSTRUMENTS:	**High Resolution Telescope, Compact Imaging Spectrometer, Context Camera, Color Imager, Climate Sounder, Shallow Subsurface Radar**

Just as the Mars Exploration Rover (see pp. 110–15) and *Deep Impact* spacecraft (see pp. 118–19) studied interplanetary space in striking new ways, NASA's *Mars Reconnaissance Orbiter* had its own unique approach. In a word, data constituted the essence of the project. Launched on August 12, 2005, from Cape Canaveral Air Force Station, Florida, aboard an Atlas V rocket, the *Orbiter* faced a long ride before it could fulfill its mission. But even five months before it entered Mars' orbit, it set a record that characterized its later role: on October 15, 2005, it returned the greatest amount of data ever sent home during an interplanetary flight. Then, on March 10, 2006, the *Orbiter* inserted itself in a path around the Red Planet and began to return startlingly clear images.

But this event only initiated the objectives of the *Orbiter*'s team at JPL. Over the following five months, controllers subjected the craft to aerobraking, a process by which they "dipped" the *Orbiter* into Mars' upper atmosphere and then lifted it out, causing it to decelerate owing to atmospheric friction, but without overheating. Gradually, the highly elongated orbit changed to an almost circular, two-hour loop quite close to the Martian surface. At the conclusion of the maneuver (September 2006), *Orbiter* came closer to the South Pole than the North, on average about 190 miles (306 km) from the surface, more than 20 percent lower than the previous generation of Mars orbiters. It also sent ten times more data per minute than any of its predecessors.

Technicians at Kennedy Space Center, Florida, prepare to encapsulate the *Mars Reconnaissance Orbiter* prior to launch.

Once the orbital mechanics had been finished, there began a twenty-five-month phase of intense scientific observation of Mars' surface, as well as its subsurface. These events occurred on a large platform. The *Mars Reconnaissance Orbiter* measured 21 ft (6.4 m) in height, 45 ft (13.7 m) in width, and came equipped with a 10-ft (3-m) dish antenna and two solar panels, each measuring about 18 x 8 ft (5.5 x 2.4 m), or 288 sq. ft (26.8 sq. m) in total, providing a minimum of 2000 W of power. It also carried a heavy load, weighing 4806 lb (2180 kg) in all, a little more than half of which consisted of propellant and pressurant.

The *Mars Reconnaissance Orbiter* reported its observations using eight instruments:

- High Resolution Imaging Science Experiment: a telescopic camera—the most powerful yet sent to a body outside of the Earth, with a 20-in. (51-cm) diameter—takes images across Mars' surface, viewing areas 3¾ miles (6 km) wide at a sweep. The *Orbiter* has three modes—targeted, regional, and global—and the high-resolution camera will be trained on targeted areas, such as dunes, gullies, and boulders. Taking pictures of subjects as small as 3 ft (91 cm) across, it can also show them in three dimensions.
- Compact Reconnaissance Imaging Spectrometer for Mars: this instrument pursues the search for evidence of water on Mars. In its regional mode, it scans for 70 wavelengths of light characteristic of minerals related to water (or its processes). Once it locates a promising site, it targets the area using 544 different wavelengths.
- Context Camera: an instrument producing black-and-white images, it covers narrow bands of the landscape (18½ miles/30 km wide) as the high-resolution camera and the imaging spectrometer see the same regions in broader swaths.
- Mars Color Imager: much like the camera flown on the failed Mars Climate Orbiter mission, the Color Imager's very wide-angle lens concentrates on the global perspective of the planet, monitoring each day's climate, as well as surface changes, increases and decreases of the polar icecaps, and the composition of clouds.
- Mars Climate Sounder: a pair of yoked telescopes that can be pointed sideways to the horizon, up into space, or down toward the planet in order to monitor nine bands of radiation and determine variations in Martian water vapor, dust, atmospheric ice, and temperature.
- Shallow Subsurface Radar: shooting radio waves into the surface over a wide region, this instrument measures the returning waves to see whether the crust contains ice, liquid water, and/or rock, to a depth of 3168 ft (966 m).
- Gravity Investigation: researchers will survey variations in the spacecraft's motions caused by gravitational force in order to infer the distribution of the planet's mass at and below its surface.
- Atmospheric Structure Investigation: during the aerobraking phase of the mission, scientists employed sensitive accelerometers to take readings of the vertical structure of the Martian upper atmosphere and thus gain a better understanding of its nature and history.

This image taken by *Mars Reconnaissance Orbiter* depicts gullies (about 830 ft/253 m wide) in a crater in the Terra Sirenum area of the planet.

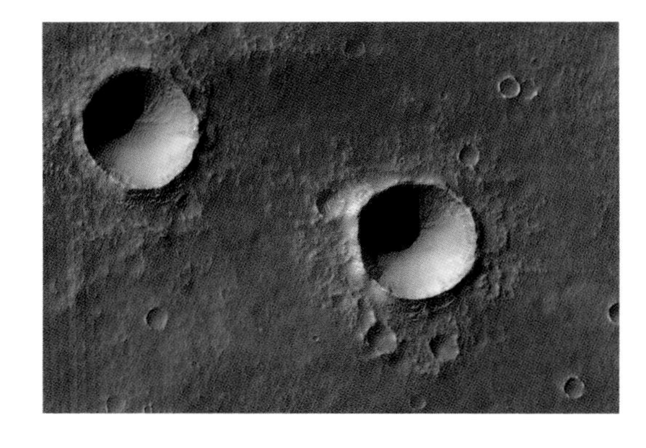

A *Mars Reconnaissance Orbiter* photograph shows two craters on the Red Planet's southern hemisphere (each roughly 2500 to 3000 ft/762 to 914 m in diameter) that have apparently experienced the same geological events.

By late spring 2007, the *Mars Reconnaissance Orbiter* had returned a staggering amount of information. In just a single year of Martian observation, it had collected enough data to fill more than 2000 compact discs.

New Horizons

Astronomers in the twenty-first century find themselves at odds over the definition of a planet, mainly as a result of the discovery of hundreds of asteroid-like objects at great distances from the Sun. As a consequence, they have decided to banish Pluto, a small sphere with a diameter of only 1400 miles (2253 km)—compared to the Earth's 8000 miles (12,874 km)—from the select company of the other eight planets. The International Astronomical Union instead considers it a "dwarf planet," under the classical definition of a sphere formed by its own gravity that rotates around the Sun.

Despite this controversy, scientists at NASA, the Johns Hopkins University Applied Physics Laboratory, and the Southwest Research Institute joined forces to mount the first mission to the planet at the end of the Solar System and to its large moon, Charon. Since its discovery in 1930 by American astronomer Clyde Tombaugh, Pluto has attracted many who have been fascinated by its unique properties. Then, during the 1990s, the fascination deepened, with the announcement that Pluto represented only the biggest and brightest of hundreds of objects circling the Sun beyond Neptune, in a region called the Kuiper Belt.

The *New Horizons* team at the Johns Hopkins University Applied Physics Laboratory conduct a "fit check" of the spacecraft's antenna dish.

After a competition in 2001 to design a spacecraft for the Pluto mission, NASA selected *New Horizons*, as proposed by the Johns Hopkins team. Fabrication of the craft was completed in 2005. Following testing at the Goddard Space Flight Center, it lifted off from Cape Canaveral, Florida, on January 19, 2006, aboard an Atlas V launcher equipped with an added Boeing third stage to boost the escape speed. Planetary mechanics dictated the launch window. Only during a three-week period at the start of 2006 could the spacecraft fly directly to Jupiter and, accelerated by 9000 mph (14,482 kmph) thanks to the gravity-assist of the mighty planet, it could reach Pluto in 2015—about three years sooner than by a direct flight. Cruising before the Jovian encounter at 41,000 mph (65,977 kmph)—the fastest speed of any spacecraft ever launched—it reached its closest point to Jupiter (1.4 million miles/2.3 million km away) on February 28, 2007.

The full set of data from the Jupiter leg of the mission reached researchers at Johns Hopkins by June 1, 2007. It included more than 700 observations of the Jovian atmosphere, rings, and closest moons (Io, Ganymede, Europa, and Callisto). The Jupiter flyby proved to be a shakedown cruise for *New Horizons*, one that it passed successfully. Relatively compact (wryly referred to as a grand piano mounted on a sports-bar satellite dish), it seemed smaller still because it had no solar panels. Because of its distance from the Sun, the spacecraft drew power instead from a nuclear-powered battery called a Radioisotope Thermoelectric Generator (RTG), converting heat from its plutonium-238 pellets into electricity. Seven pieces of equipment were used to collect the Jupiter data, and will later be used on Pluto and the Kuiper Belt:

- Long Range Reconnaissance Imager: a visible-light telescope that acts as a digital camera with a large telephoto lens, capable of withstanding Pluto's harsh environment.
- A Lightweight Imaging Spectrometer for Cometary Exploration (ALICE): a machine designed to probe Pluto's atmosphere, which concentrates on ultraviolet light in the far and extreme ranges.
- Ralph Telescope: a visible-light channel and a near-infrared imaging spectrometer that constitutes the main eye of *New Horizons*, it will map Pluto, Charon, and Kuiper Belt objects.
- Solar Wind Analyzer Around Pluto (SWAP): an instrument that measures the interchange between Pluto and the solar wind.
- Pluto Energetic Particle Spectrometer Science Investigation (PEPPSI): a device designed to record the escape of neutral atoms from Pluto's atmosphere.
- Student Dust Counter: an apparatus used to detect dust grains encountered by *New Horizons*, conceived and built by students at the University of Colorado, Boulder.
- Radio Science Experiment (REX): a small printed circuit board that detects powerful radio signals sent from NASA's Deep Space Network on the Earth and analyzes Pluto's and Charon's atmospheres as the waves pass through them.

Key facts

LAUNCHED:	January 2006
ESTIMATED ARRIVAL AT PLUTO:	2015
MANAGEMENT:	NASA Goddard Space Flight Center, Johns Hopkins University, Southwest Research Institute
JUPITER FLYBY:	February 2007
PURPOSE:	Survey Pluto, its moon, Charon, and the Kuiper Belt
INSTRUMENTS:	Visible light telescope, Lightweight Imaging Spectrometer, Ralph Telescope, Solar Wind Analyzer, Pluto Energetic Particle Spectrometer Science Investigation, Student Dust Counter

New Horizons's Long Range Reconnaissance Imager captured this composite picture of Jupiter, assembled at the Johns Hopkins Applied Physics Laboratory.

Jupiter's moon Ganymede, one of many images of Jupiter's system taken by *New Horizons*.

New Horizons detected a volcanic plume near the North Pole of Jupiter's moon Io, 180 miles (290 km) high.

TOP
An artist imagines what *New Horizons* might look like during its approach to Pluto and its three moons in summer 2015.

ABOVE
An illustration of *New Horizons* as it makes its descent to Pluto, with the moon Charon in the distance.

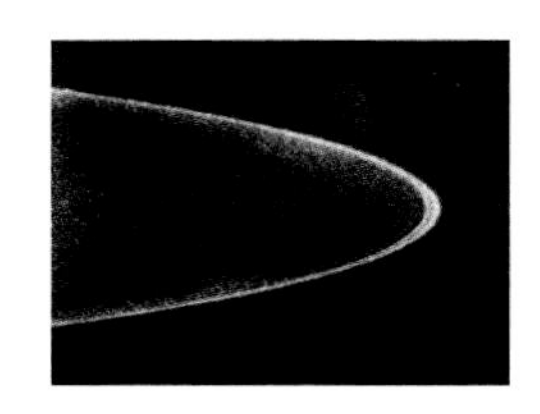

On February 24, 2007, *New Horizons* took a picture of Jupiter's ring system from a distance of 4.4 million miles (7 million km).

Mars Odyssey

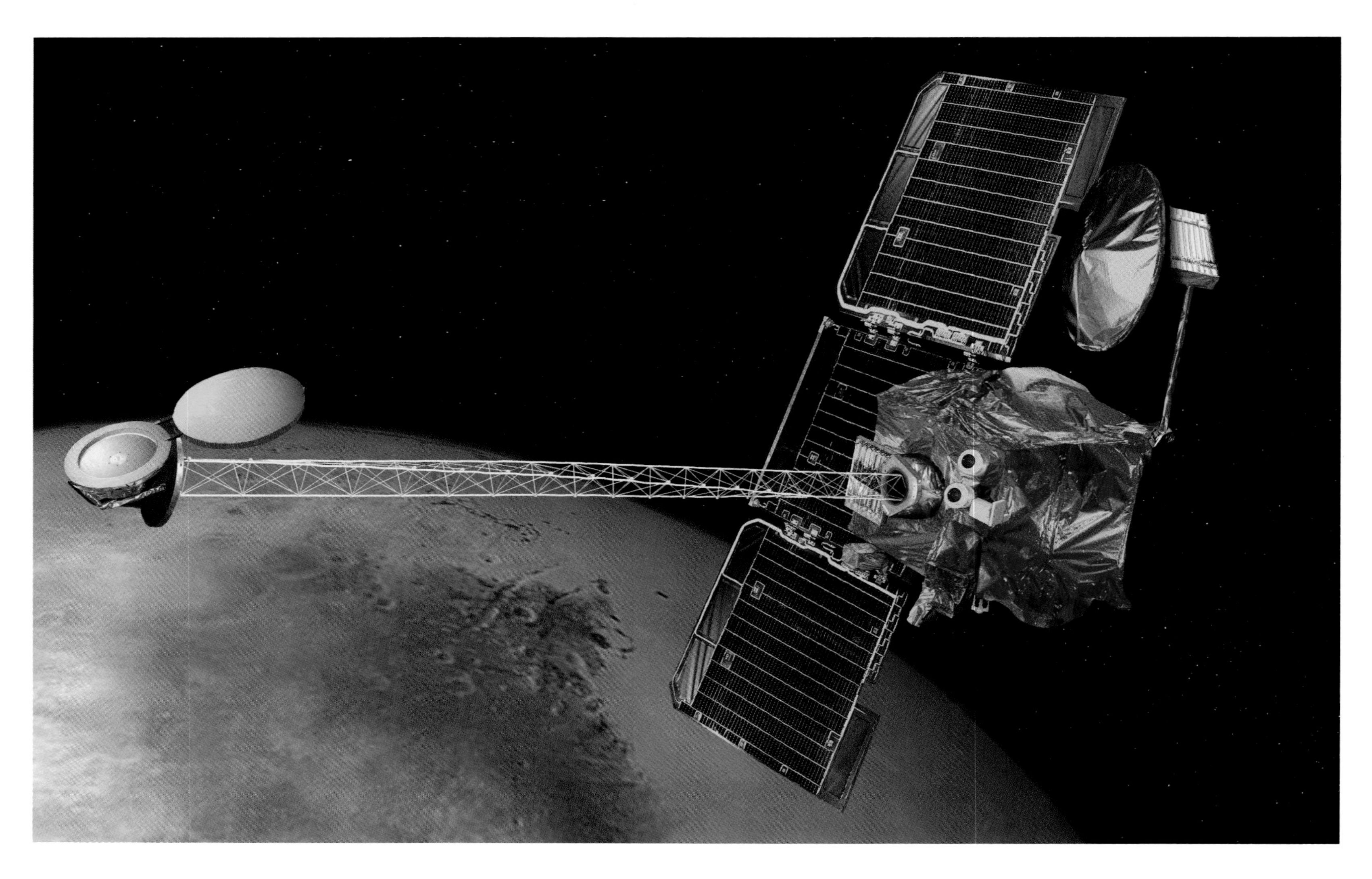

Two drawings of *Mars Odyssey* making observations above the Red Planet.

In addition to conducting in-house interplanetary research, NASA scientists also collaborate with counterparts in other parts of the world. Just as Russia and the United States entered a period of intense collaboration during the preparation for, and construction of, the International Space Station (see pp. 26–29), the two countries have extended their partnership into the exploration of the Solar System. In this instance, the Red Planet once again proved to be the most alluring of space objects, embodied in the *Mars Odyssey* project.

The Jet Propulsion Laboratory initiated and managed *Mars Odyssey* and contracted Lockheed Martin Space Systems of Denver, Colorado, to fabricate the spacecraft. *Mars Odyssey* measured just 7 ft 2 in. (2.2 m) in length, 5 ft 7 in. (1.7 m) in height and 8 ft 6 in. (2.6 m) in width, and carried a solar array 18 ft 8 in. (5.7 m) long. It weighed a mere 1609 lb (730 kg), and only 809 lb (367 kg) without propellant. A Delta II launch vehicle hoisted the spacecraft out of the atmosphere on April 7, 2001, from Cape Canaveral, Florida. It reached the Red Planet on October 24, and through the process of aerobraking (which saved some 440 lb/200 kg of propellant needed for the thrusters) slowed down

Key facts

MANAGEMENT:	**NASA Jet Propulsion Laboratory (with Russian contribution)**
MANUFACTURER:	**Lockheed Martin Space Systems**
LAUNCHED:	**April 2001**
MAPPING OF MARS:	**Initiated February 2002**
ANCILLARY ROLE:	**Communications relay for Mars Rover project**
PURPOSE:	**Survey elements on and below the Martian surface associated with water and ice**
INSTRUMENTS:	**High Energy Neutron Detector, Gamma Ray Spectrometer, Thermal Emission Imaging System**

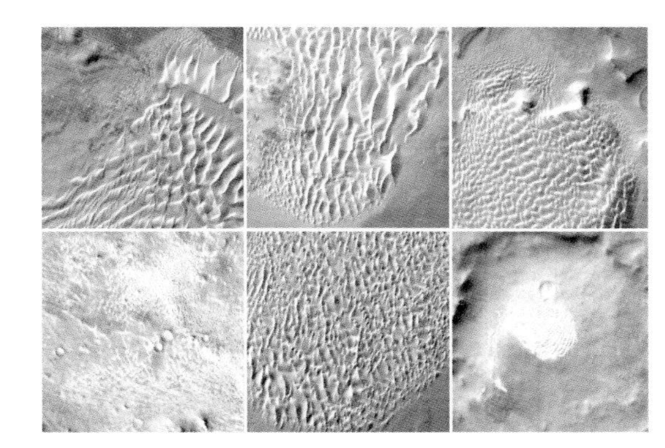

ABOVE
In addition to acting as a communications relay between the Mars Exploration Rovers and ground controllers, *Mars Odyssey* captured many breathtaking images of the Red Planet. This color view shows the Solar System's biggest canyon, with layered slopes and basalt dunes.

RIGHT
***Mars Odyssey*'s infrared imaging is applied here to six different Martian dunes.**

and assumed a more circular orbit, so that by February 2002 it could begin its mapping objectives. The primary science objectives of *Mars Odyssey* ended in August 2004, and it then began an extended mission, and a second one beyond that. Meanwhile, since January 2004 it had been performing an important ancillary function as communications relay for the Mars Exploration Rovers *Spirit* and *Opportunity*.

The Russian Federal Space Agency (Roskosmos) made a pivotal contribution to *Mars Odyssey*. To further the continuing search for water on Mars, Dr. Igor Mitrofanov of the Russian Space Institute led a team that created a device called the High Energy Neutron Detector (HEND), a key part of the spacecraft's Gamma Ray Spectrometer and one of two instruments for detecting neutron emissions (the other was devised by the Los Alamos National Laboratory in New Mexico). HEND peers down at the Martian surface and penetrates the iron-mineral-bearing dust and dirt looking for elements characteristic of subsurface water. To date, it has actually succeeded in detecting buried ice in the Martian soil. In addition, the Russian scientists—whose space program commonly produces a second instrument identical to the one actually flying—proposed bringing the twin to the International Space Station to monitor neutrons (solar radiation) above the Earth's atmosphere. Russian cosmonaut Mikhail Tyurin and astronaut Michael Lopez-Alegria installed it in February 2007. To expand the spirit of international cooperation on the project, the JPL and Russian Space Institute teams met occasionally—sometimes in Moscow, sometimes in America—to analyze the data and to prepare for future collaborations.

Three instruments flew aboard the *Mars Odyssey*:

- The Gamma Ray Spectrometer measures the gamma rays emitted by various chemical elements on the planet in response to bombardment by cosmic rays. This enables researchers to estimate the abundance of different elements, as well as their distribution. Furthermore, by measuring neutrons, the researchers are able to calculate the presence of hydrogen just below the surface, indicating water or ice.
- The Thermal Emission Imaging System scans in the visible and infrared spectra at the surface, looking for minerals that typically form in water—carbonates, silicates, oxides, sulfates, and others. To date, researchers have found that ice on the ground varies in depth to a greater degree than previously expected.
- The Martian Radiation Environment Experiment uses a spectrometer to detect the extent of radiation emanating as energetic particles from the Sun, and as cosmic rays from outside the Solar System. Both pose potentially grave health risks to space travelers.

Cassini–Huygens

BELOW
A diagram of *Cassini–Huygens*, illustrating the *Huygens* probe (far side) mounted to *Cassini*.

ABOVE
An artist's version of *Cassini* above Saturn's largest moon, Titan, with *Huygens* flying independently and Saturn nearby.

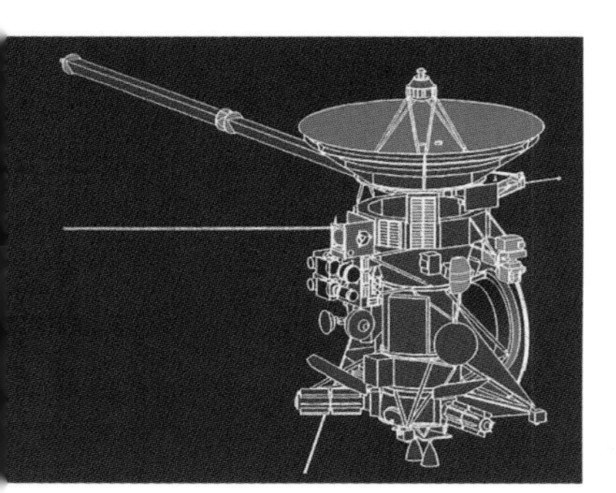

Key facts

MANAGEMENT:	**NASA Jet Propulsion Laboratory (*Cassini*), ESA and the Italian Space Agency (*Huygens*)**
LAUNCHED:	**October 1997**
ARRIVED AT SATURN:	**July 2004**
SUCCESSFUL DEPLOYMENT OF THE HUYGENS PROBE ON TITAN:	**January 2005**
PURPOSE:	**Explore Saturn, its rings, and its eight main moons**

An expansive international planetary endeavor occurred when NASA, the European Space Agency (ESA), the Italian Space Agency (ASI), and seventeen contributing countries combined their efforts in the ambitious *Cassini–Huygens* mission. Named for Dutch scientist Christiaan Huygens (1629–1695) and the Italian astronomer Jean-Dominique Cassini (1625–1712)—who, respectively, discovered and confirmed the existence of Saturn's rings—the spacecraft consisted of two parts: an orbiter bearing the name *Cassini*, and a landing probe known as *Huygens*. Fittingly, its mission centered on Saturn: to explore the Saturnian system, including the planet itself, its rings, and its eight major moons. Managing the overall program, the JPL conceived, designed, and fabricated the *Cassini* orbiter. The *Huygens* probe underwent development by the European Space Technology and Research Center, with Aérospatiale of Cannes, France, acting as the prime contractor. ESA supervised *Huygens*'s operations through the ESA control center in Darmstadt, Germany. ASI contributed several of *Cassini*'s science instruments, much of its radio system, as well as the spacecraft's high-gain antenna.

The *Cassini–Huygens* duo needed heavy launch capacity. Mounted atop a 2.2 million-lb (998,000-kg) Titan IVB–Centaur combination, it lifted off from Cape Canaveral, Florida, on October 15, 1997, beginning an odyssey of more than 1 billion miles (1.6 billion km), which lasted almost seven years. *Cassini* and *Huygens* together weighed a substantial 12,593 lb (5712 kg) with fuel, and stood 22 ft (6.7 m) high and 13 ft (4 m) wide. The circular, shellfish-shaped *Huygens* measured almost 9 ft (2.7 m) in diameter. Upon leaving the Earth, *Cassini–Huygens* picked up speed as it flew by Venus in April 1998 and June 1999; passed the Earth in August 1999 and Jupiter in December 2000; and arrived at Saturn in July 2004, at which time an onboard rocket engine fired to brake the spacecraft's speed and drop it into orbit. Some seventy-five orbits would follow in the four-year mission.

Meanwhile, the time for *Huygens* had come. In late 2004, *Cassini* altered its orbit and released the cone-like *Huygens* on a three-week, unpowered, autonomous voyage to Saturn's largest moon, Titan, traveling at 13,400 mph (21,563 kmph) as it encountered the moon's upper atmosphere. Then, two-and-a-half hours from its target, three sets of parachutes opened and *Huygens* made a slow descent through Titan's hazy and nitrogen-filled atmosphere, to land safely on the moon on January 7, 2005. For seventy minutes it continued sending data to the *Cassini* orbiter (which relayed the signal to Earth), until finally the stream stopped. In the meantime, during the descent and landing, six instruments enabled scientists to capture more than one thousand images from its descent imager, sample Titan's thick atmosphere from top to bottom, measure winds and temperatures, and see through the haze and clouds to map its surface.

ABOVE
An artist studied the images received from *Huygens* after it landed, and painted this picture of the probe resting on Titan's landscape.

LEFT
Two technicians converse as they integrate the *Huygens* probe with its heat shield.

RIGHT
The *Huygens* probe mated to the back of the erect *Cassini* spacecraft prior to their seven-year journey to Saturn and its moons.

ABOVE

At about a 40-degree angle from Saturn's ring plane, *Cassini* took thirty-six images of the planet, integrated into this single, startling picture.

The *Cassini* spacecraft, chartered (among many other things) to make a detailed study of Saturn's rings, reveals some of its findings in this image of intricate patterns at the outermost part of the rings' spoke-forming region.

Cassini's camera takes in a wondrous sight. From this perspective, the rings of Saturn seem to bisect its distant moon Titan.

Cassini witnesses the moon Titan awash in a 360-degree sunset.

Titan, as seen by *Cassini*, in a composite image made using the infrared and ultraviolet wavelengths.

Before and after *Huygens*'s brilliant success, *Cassini* continued to do its work. Even before its first orbit of Saturn, it passed the moon Phoebe in June 2004, then with each circle around the giant planet it flew by moons at distances as close as 311 miles (500 km) and as far away as more than 40,000 miles (64,374 km). It will fly by Titan forty-five times, Enceladus four times, Iapetus twice, Mimas once, Rhea twice, Hyperion once, Dione once, and Tethys twice. Its twelve powerful instruments will study the composition of Saturn's atmosphere, as well as its clouds, winds, waves, and eddies; do much the same for Titan; investigate the magnetosphere of the ringed planet, as well as the rings themselves, discovering more about their composition, structure, and dynamic processes, and the interrelationships between Saturn's moons and rings; and discover the compositions and geologic histories of Saturn's "icy" moons. At the same time, the rich vein of data mined by *Huygens* will continue to be assayed.

***Cassini* captures the southern hemisphere of Saturn, revealing a great hurricane-like storm over the pole.**

Another finely detailed *Cassini* portrait, this one of the spoke-forming area of Saturn's rings.

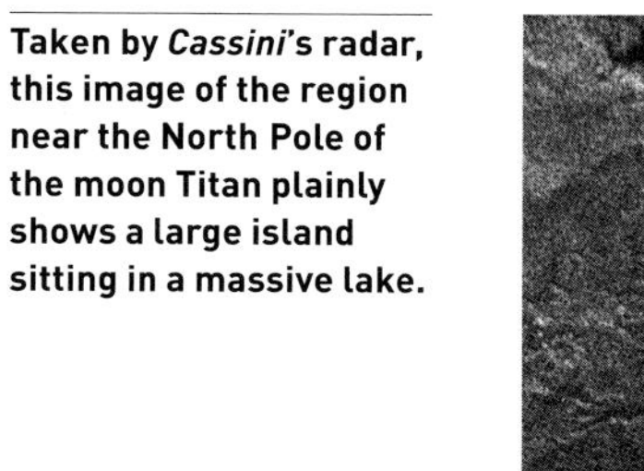

Taken by *Cassini*'s radar, this image of the region near the North Pole of the moon Titan plainly shows a large island sitting in a massive lake.

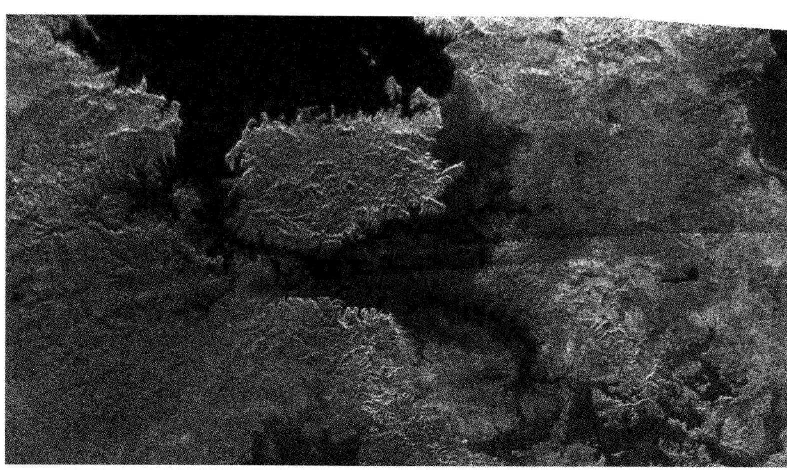

ESA's *Huygens* probe captured this image of the surface of Titan on January 14, 2005.

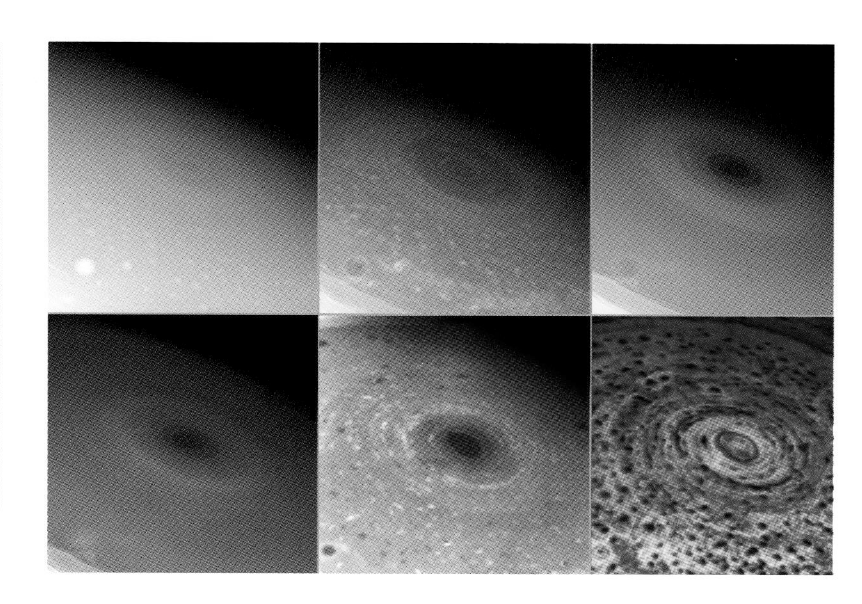

***Cassini*'s six images of Saturn's South Pole reveal a hurricane-like storm, the rotating motion of which is shown at varying depths of cloud cover.**

Mars Express

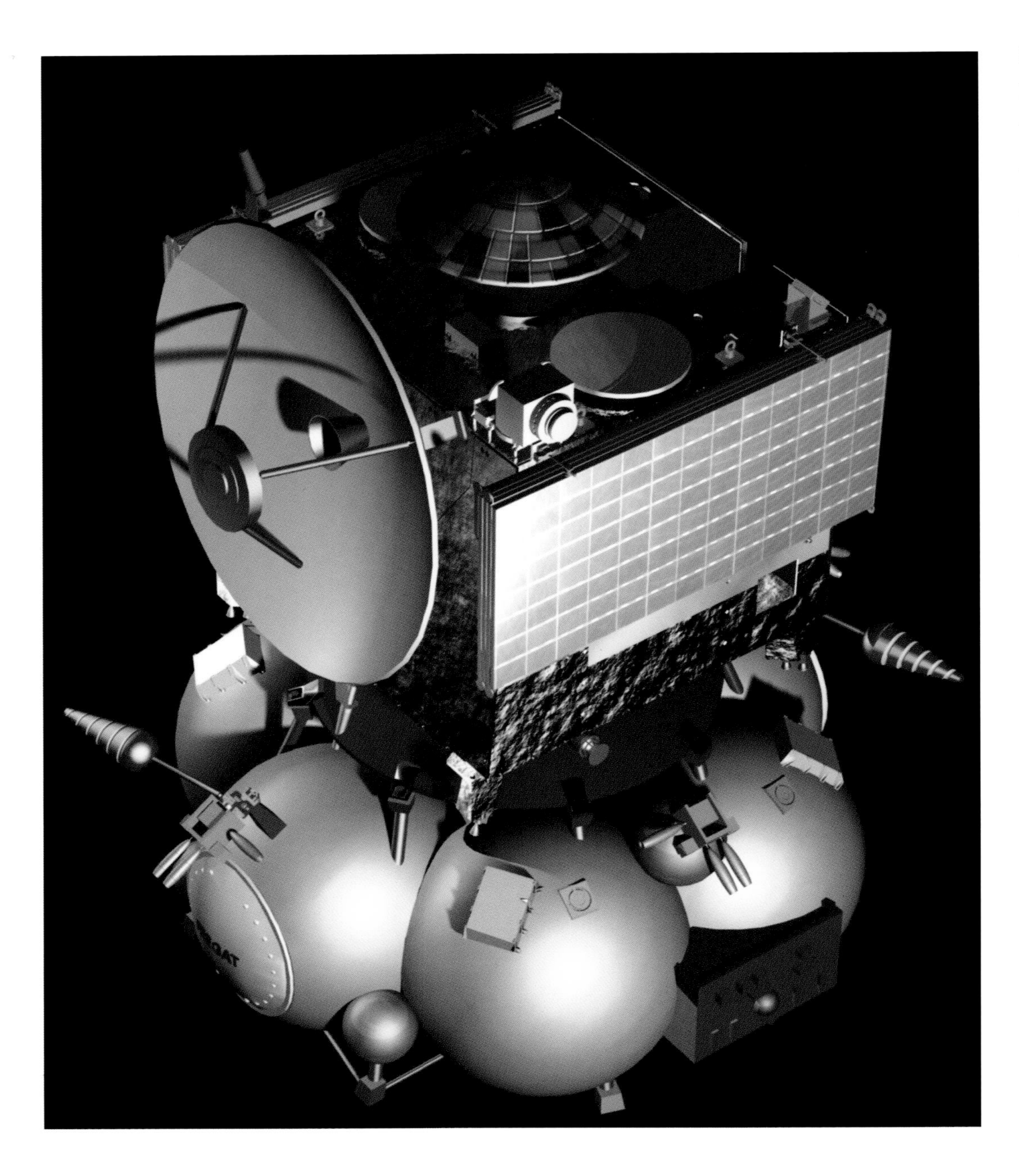

Key facts

MANAGEMENT:	**ESA**
MANUFACTURER:	**Astrium SAS**
LAUNCHED:	**June 2003**
ESTIMATED MISSION END:	**May 2009**
PURPOSE:	**Land a probe on Mars (Beagle 2), while the orbiter (Cassini) analyzes volcanic, glacial, and water activity around the planet**
FAILURE OF BEAGLE 2:	**December 2003**
MARS EXPRESS INSTRUMENTS:	**High Resolution Stereo Camera, OMEGA Spectrometer, SPICAM Spectrometer, Planetary Fourier Spectrometer, ASPERA, and MARSIS**

Mars Express mounted on the Fregat booster, the fourth stage of the Soyuz–Fregat rocket. The Fregat separated from _Mars Express_ in Earth orbit, after placing it on a trajectory to Mars.

An artist's rendering of _Mars Express_ orbiting the Red Planet after a six-month journey from Earth.

Not long after NASA, ESA, and the other nations combined forces to launch the spectacularly successful *Cassini-Huygens* mission (see pp.126–28), ESA initiated its own series of planetary probes. *Mars Express* represented perhaps the boldest of the European attempts because, like *Cassini-Huygens*, it held the promise of a long-term orbiter, in addition to an autonomous landing probe.

Mars Express's designers approached the project with the ambitious intention of reducing costs while producing the first all-European spacecraft to visit Mars. In the end, ESA spent about one-third as much on this mission as it had on others of equal scope. An industry competition in 1998 resulted in the naming of Astrium SAS of Toulouse, France, as the prime contractor. After the development contract had been signed in March 1999, project engineers began their work, borrowing from at least two sources: from the Russian *Mars 96* spacecraft, as well as from ESA's contemporary *Rosetta* mission (see pp. 136–37). In addition, Astrium purchased launch support aboard the Russian Soyuz-Fregat launch system, built by Starsem (an alliance between Arianespace, Aerospatiale, the Russian Aviation and Space Agency, and the Samara Space Center). In all, Astrium led a group of twenty-four firms in fifteen European nations, as well as American interests.

About four years after the contract signing (relatively fast in order to take advantage of a planetary alignment), *Mars Express* lifted off from Baikonur Cosmodrome in Kazakhstan on June 2, 2003. Six months later, *Mars Express* drew near to the Red Planet. Five days before its arrival, on December 19, 2003, it ejected a small Martian lander known as *Beagle 2*, after the famous ship that carried Charles Darwin on some of his voyages of discovery. The release and descent presumably occurred as intended, with a parachute opening to brake the fall and gas-filled bags inflating as *Beagle 2* bounced to a halt. But none of this can be certain. As *Beagle 2* descended, mission controllers lost contact with the craft, and despite efforts by other Mars orbiters and radio telescopes on the Earth to detect signals, none could be picked up. The loss to the mission, while not devastating, certainly had real consequences, preventing "hands-on" discoveries on the surface itself, especially rock and soil analyses.

Yet the orbiter portion of *Mars Express* functioned well. Despite its small size (a box roughly 5 x 6 x 5 ft/1.5 x 1.8 x 1.5 m) and light weight (2470 lb/1120 kg at launch, including the 132-lb/60-kg *Beagle 2*), its seven instruments—High Resolution Stereo Camera, OMEGA Visible and Infrared Mineralogical Mapping Spectrometer, SPICAM Ultraviolet and Infrared Atmospheric Spectrometer, Planetary Fourier Spectrometer, ASPERA Energetic Neutral Atoms Analyzer, Mars Radio Science Experiment, and MARSIS Sub-Surface Sounding Radar Altimeter—have sent home years' worth of valuable data.

Mars Express has uncovered evidence of fluvial, glacial, and volcanic activity from earliest times until recently; picked up traces of methane, suggesting life in the present, or at least active vulcanism; and discovered mineralogical evidence on the surface that enables researchers to say with confidence that the planet had abundant water in earlier times. Indeed, between 2004 and 2006, *Mars Express* performed so well that ESA decided to extend its life until May 2009. Meanwhile, the spacecraft will continue to map the surface in unique, high-resolution, stereoscopic color images, and to plumb beneath the surface of the planet for water and ice.

Taken in an area of Mars known as Aureum Chaos, this *Mars Express* image reveals a bright substance (bottom, right) consistent with hydrothermal activity or the evaporation of fluids.

Taken near the Martian equator by *Mars Express*, this picture appears to show a frozen sea covered in dust.

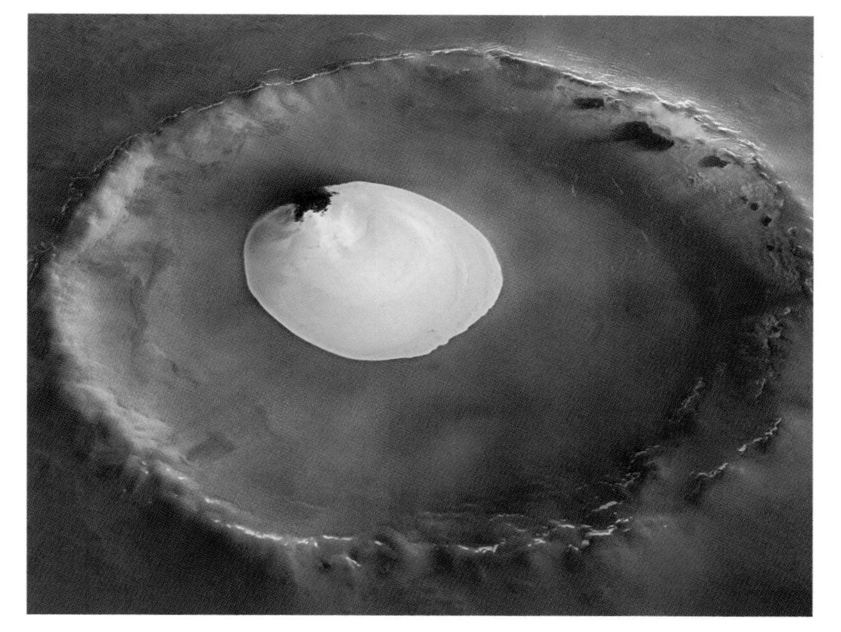

ABOVE
Residual water ice, located in the Vastitas Borealis Crater, in an image taken by *Mars Express*.

PAGES 132–33
A transparent image of *Mars Express* as seen from above.

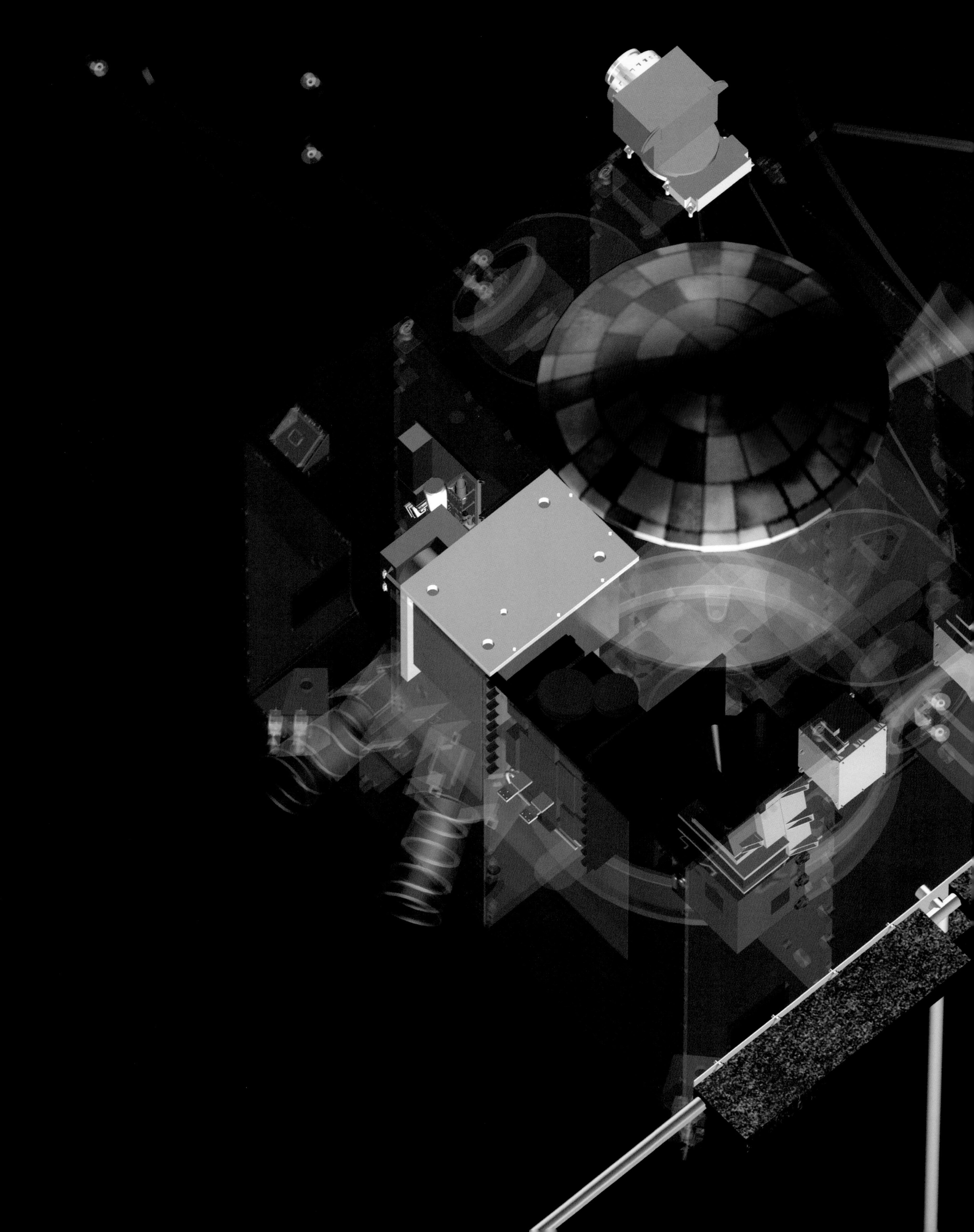

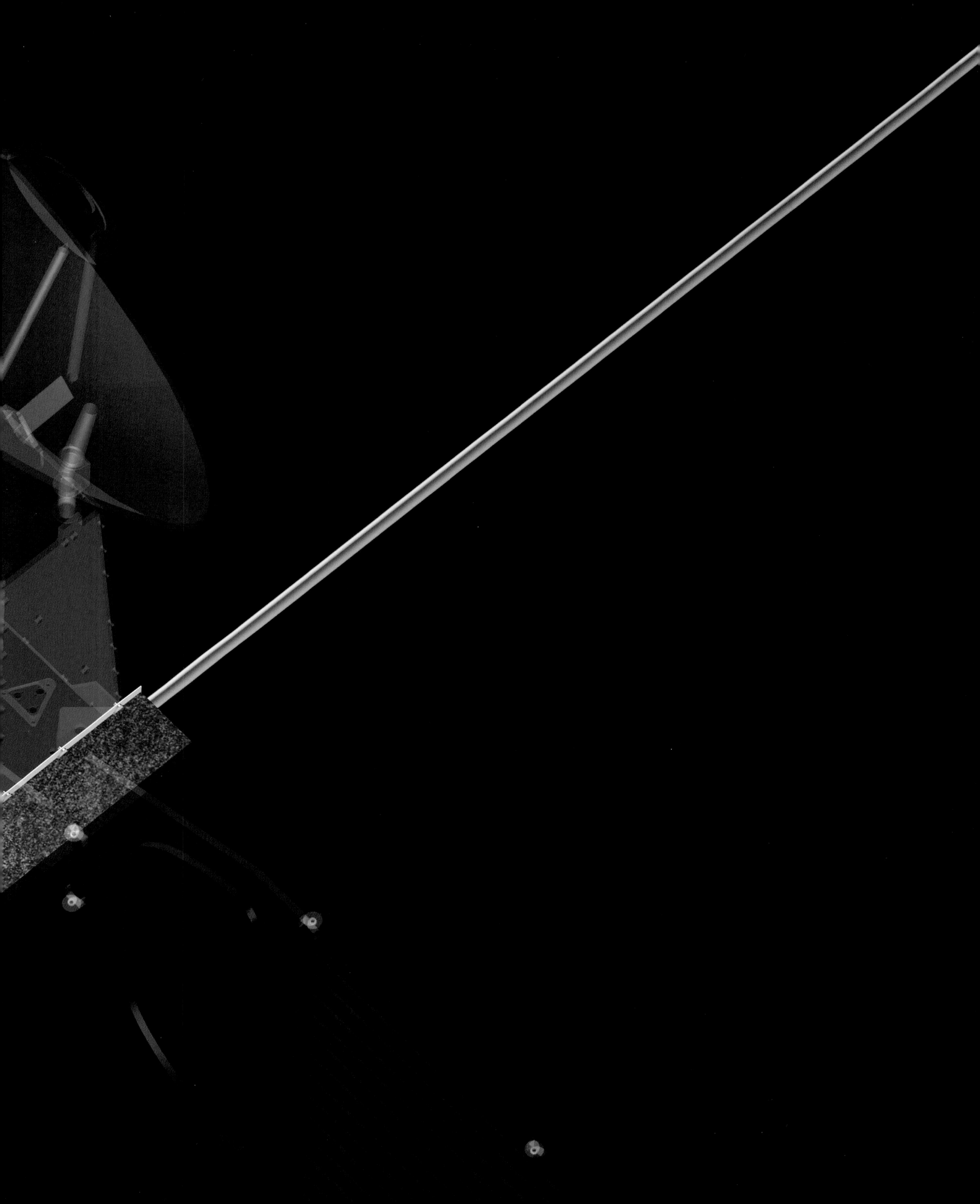

Venus Express

An artist imagines the *Venus Express*'s main engine burn, which decelerates the spacecraft so that it can achieve orbital insertion.

Key facts

MANAGEMENT:	**ESA**
MANUFACTURER:	**Astrium SAS**
LAUNCHED:	**November 2005**
BEGAN TO ORBIT VENUS:	**April 2006**
ESTIMATED MISSION END:	**May 2009**
PURPOSE:	**Observe the unique, hot, thick, and super-rotating Venutian atmosphere**

Venus Express, very similar to its brother *Mars Express* (see pp. 130–33) instead circled the Hothouse Planet. Overall, the design and components did not vary between the two spacecraft, and the prime contractor and the launch arrangements remained the same. But ESA did make some changes, mainly in recognition of the differences in Venutian versus Martian planetary characteristics. Because Venus orbits the Sun at half the distance of Mars, *Venus Express*—exposed to four times the heat—required thermal protection measures. Its radiators are bigger and more efficient. The multilayer insulation not only has greater thickness, but is also wrapped differently than on *Mars Express*. *Venus Express* bears a gold coat to reflect heat; *Mars Express* has a black exterior to absorb it. Moreover, at about 26 ft (8 m) in length, *Venus Express*'s gallium arsenide-based solar arrays measure half of those on the silicone-based *Mars Express* because of the greater intensity of the Sun at Venus' position in the Solar System. Furthermore, pitted against a gravity eight times that of Mars (and much like that of the Earth), *Venus Express* required 20 percent more propellant for its thrusters to have the fuel necessary to brake the spacecraft. Finally, *Venus Express* had no lander.

Venus Express blasted off from Baikonur Cosmodrome in Kazakhstan on November 9, 2005. On April 11 of the following year, the spacecraft's main engine fired for fifty minutes, initiating its orbit around Venus. A month later, its orbit had been changed from a highly elliptical nine-day circle to a polar one lasting twenty-four hours. In so doing, it began to observe one of the most unusual atmospheres in the Solar System. Subject to extremes of pressure, Venus' very hot, thick carbon dioxide atmosphere acts like a greenhouse, spinning around the planet in a super-rotation that lasts just four days. *Venus Express* also had the opportunity to make the first optical observations of the surface of the planet, using a select channel in the infrared spectrum to see through the clouds. In addition, the spacecraft was tasked with searching for volcanic and tectonic activity.

After ten months of successful operations—just halfway through its planned mission—ESA decided in February 2007 to extend *Venus Express*'s service to early May 2009. This new lease of life made some important work possible. Until that point, the spacecraft had astounded researchers with its images, including those of a massive, double-eyed atmospheric vortex at Venus's south pole, which provided an insight into the workings of the planet's atmosphere. Then, in the last week of May and the first week of June 2007, thirteen Earth-based telescopes around the world trained their gaze on Venus in order to complement data streaming in from the spacecraft, thus enriching and deepening its observations.

Finally, even as these sightings went on, NASA's *MESSENGER* spacecraft (see pp. 116–17)—which was on its way to Mercury—approached during its second flyby of Venus, affording the opportunity for NASA and ESA to collaborate in a planetary viewing on the fly. On June 6, 2007, for a few hours, *MESSENGER* and *Venus Express* watched the planet almost simultaneously, offering two points of reference for the same objects, with very little lapse in time. Meanwhile, ground instruments on the Earth augmented the experiment. For instance, as *MESSENGER* and *Venus Express* collected data about the dynamics of the Venutian atmosphere, the Observatoire de Haute-Provence in France took real-time readings of the planet's winds, adding depth and dimension to the project as a whole.

Venus Express being prepared for shipment from Toulouse, France, to the launch site in Kazakhstan.

Two technicians in Toulouse stand by the _Venus Express_, voyager to the Earth's closest neighbor in the Solar System.

Venus Express took this picture of Venus on its first orbit using an ultraviolet/visible/near infrared spectrometer.

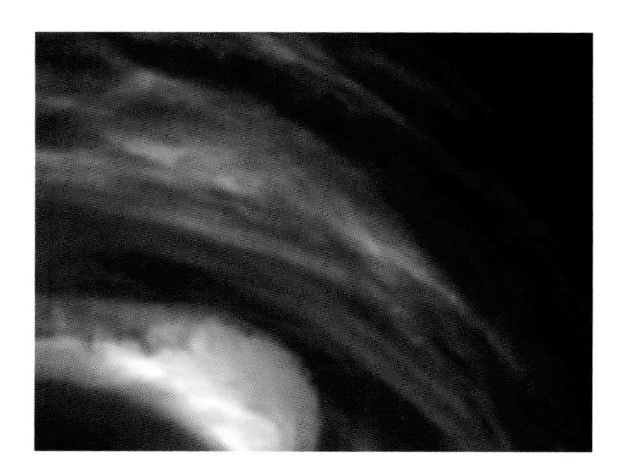

This image—taken on September 23, 2006—shows a portion of Venus near its South Pole, viewed from the most distant point in _Venus Express_'s orbit.

Rosetta

Key facts

MANAGEMENT:	**ESA**
LAUNCHED:	**March 2004**
ANTICIPATED ARRIVAL AT COMET 67 P / CHURYUMOV-GERASIMENKO:	**Between January and May 2014**
ANTICIPATED INTERCEPTION OF AND LANDING ON THE COMET:	**November 2014**
PURPOSE:	**Place a lander on a distant comet to discover its origins, while an orbiter maps it from above**

ABOVE
An artist's conception of *Rosetta*, its lander, and the mission's target, Comet 67P/Churyumov-Gerasimenko.

ABOVE, RIGHT
A drawing of *Rosetta* on Comet 67P/Churyumov-Gerasimenko, where it will operate for a minimum of sixty-five hours.

RIGHT
***Rosetta* undergoes vibration tests at ESA's test center in The Netherlands.**

FAR RIGHT
Thermal testing on *Rosetta* in the Large Space Simulator at ESA's European Space and Technology Center.

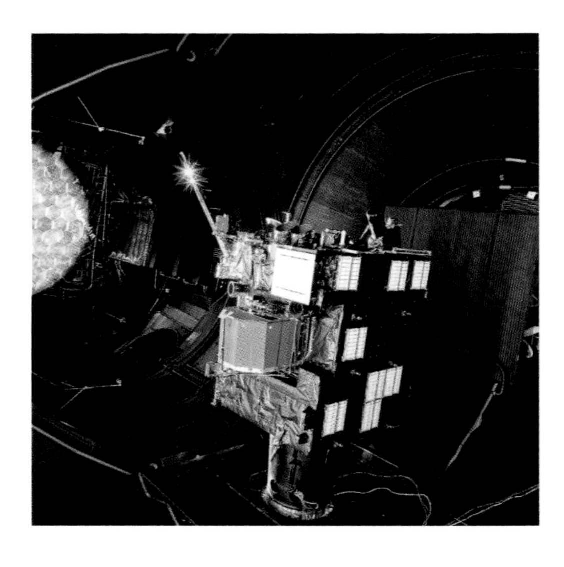

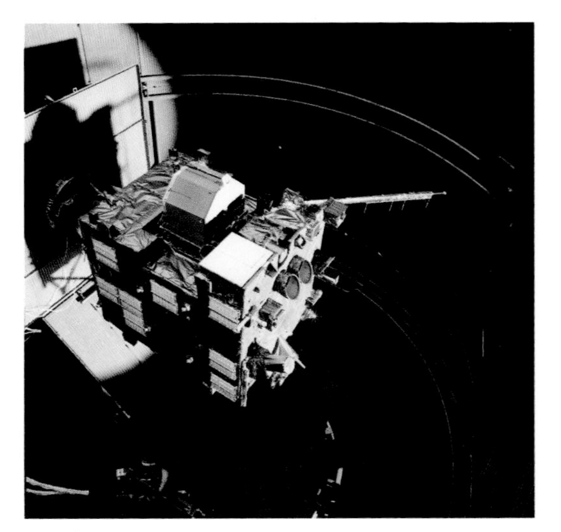

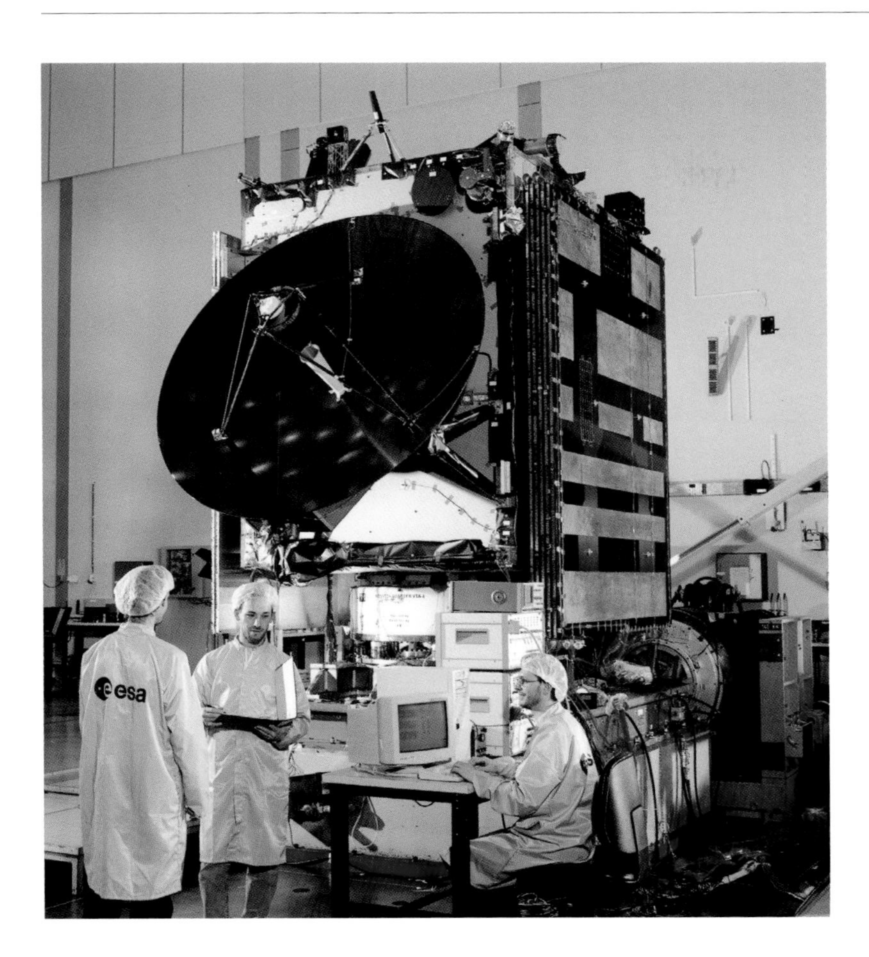

LEFT
Rosetta—ESA's comet explorer—underwent inspection of its solar panels at the space agency's test facilities in Noordwijk, The Netherlands.

RIGHT, ABOVE
A composite image of Mars, taken during a *Rosetta* flyby; it is not in true color, but clearly shows the spring polar ice cap.

RIGHT, BELOW
A true-color image of the Red Planet, with a clear depiction of the South Pole's ice cap.

Like NASA's *Deep Impact* before it (see pp. 118–19), Europe's *Rosetta* focuses on the life of comets. Its planners wanted it, unlike *Deep Impact*, to land on, not blast into, a comet. *Rosetta*, like the triumphant NASA/ESA *Cassini-Huygens* (see pp. 126–29), is a combined orbiter-lander, and the *Rosetta* team hoped for equal success. Indeed, the spacecraft bore this wish in its name, which calls to mind the famous slab discovered by French soldiers in 1799 in Rashid (Rosetta), Egypt. Carvers from ancient Egypt inscribed the same text on this piece of basalt in hieroglyphics, demotic, and Greek, enabling scholars to decipher the writing of the ancient pharaohs. Just as the Rosetta Stone had broken the linguistic code of the Egyptians, modern contributors to the *Rosetta* spacecraft hoped their work would help decode the origins of the planets by studying a body that predated them, known as Comet 67 P/Churyumov-Gerasimenko.

The voyage of *Rosetta* began on March 2, 2004, when it rose from the pad at ESA's launch site in Kourou, French Guiana, riding on an Ariane 5 rocket. Somewhat like the American *New Horizons* project (see pp. 122–23), *Rosetta* will make progress slowly. Powerful as the Ariane 5 may be, the approximately 6600-lb (3000-kg) *Rosetta* weighed too much and needed to travel too far to be propelled directly to its target by rocket power alone. Instead, in order to pick up sufficient speed to reach Churyumov-Gerasimenko, it needed to make flybys of the Earth in 2005 and 2007 (with one more planned for 2009), and, in February 2007, approach Mars. *Rosetta* will intercept Churyumov-Gerasimenko, which has an almost 2½-mile (4-km) nucleus and circles the Sun every 6½ years, about ten years after being launched, between January and May 2014, at the point in the comet's orbit at which it is farthest from the Sun. *Rosetta* will then begin mapping the comet in August of that year, release a lander in November, and continue to circle Churyumov-Gerasimenko for just over a year, from November 2014 until December 2015.

Rosetta not only has considerable weight, but also significant size, its orbiter consisting of a box roughly 9 x 7 x 7 ft (2.7 x 2.1 x 2.1 m) and big solar arrays, some 105 ft (32 m) from end to end. It has a large communications dish mounted on one side and the lander on the other. Upon release from *Rosetta*, the small, box-shaped lander—weighing about 221 lb (100 kg)—will open its three legs and head for the comet in a ballistic descent. The legs will absorb the shock during touchdown, after which a harpoon will be launched into the surface to pin the spacecraft to the comet, a measure necessary because of the comet's very weak gravity. The lander might communicate with the orbiter for as short a time as a week, or for as long as several months.

Intended to measure almost every conceivable feature above, on, and under the surface of Churyumov-Gerasimenko, *Rosetta* was packed with no fewer than eleven instruments on board the orbiter and nine on the lander. The orbiter held an Ultraviolet Imaging Spectrometer, as well as a Comet Nucleus Sounder, a Cometary Secondary Ion Mass Analyzer, a Grain Impact Analyzer and Dust Accumulator, a Micro-Imaging Analysis System, a Microwave Instrument, an Imaging System, a Spectrometer for Ion and Neutral Analysis, a Plasma Consortium Experiment, a Radio Science Investigation, and a Visible and Infrared Mapping Spectrometer. The lander was equipped with an Alpha Proton X-Ray Spectrometer, an Imaging System, a Comet Nucleus Sounding System, a Cometary Sampling and Composition Experiment, an Evolved Gas Analyzer, a Multi-Purpose Sensor for Surface and Subsurface Science, a RoLand Magnetometer and Plasma Monitor, a Sample and Distribution Device, and a Surface Electrical and Acoustic Monitoring Experiment.

Hayabusa

BELOW
An illustration of the *Hayabusa* spacecraft in flight.

RIGHT
An artist's impression of a hovering *Hayabusa* attempting to collect samples from the asteroid Itokawa.

BELOW, RIGHT
The *Hayabusa* spacecraft depicted just above Itokawa, with the *Minerva* lander on the surface.

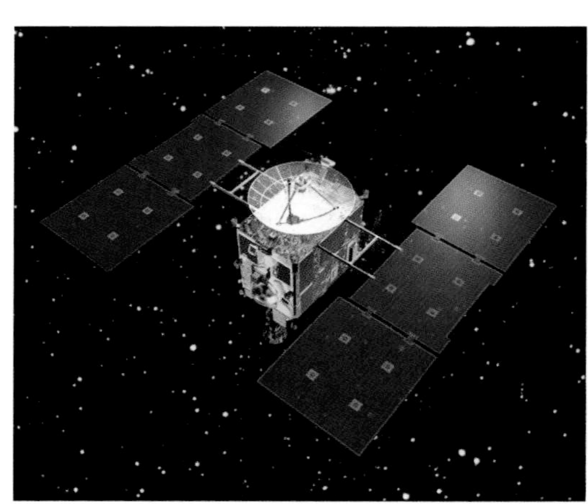

Key facts

LAUNCHED:	**May 2003**
REACHED ASTEROID ITOKAWA:	**September 2005**
HAYABUSA GOES SILENT:	**December 2005 to March 2006**
CONTROLLERS SEND HAYABUSA BACK TO EARTH:	**April 2007**
ESTIMATED LANDING:	**2010**
PURPOSE:	**Collect samples from a distant asteroid and return them to Earth**

The object of the *Hayabusa* spacecraft's many travails, the asteroid Itokawa, photographed by *Hayabusa*.

Reminiscent of the earliest reconnaissance satellites, which had to return photographic film to Earth for processing, the *Hayabusa* (Japanese for "falcon") spacecraft has attempted to send home space samples, but under far more complicated circumstances. The *Hayabusa* team chose this spacecraft to demonstrate four path-breaking technologies, any one of which would constitute a breakthrough: electric propulsion, autonomous navigation, sample collection from an asteroid, and safe return of the specimens. Electric power alone had great promise for future planetary missions due to high efficiency. On *Hayabusa*, the electric engine is operated by ionizing a propellant (xenon) using microwaves, accelerating the ions in a powerful electric field, then releasing them at high speed to cause thrust.

Launched from Kagoshima Space Center in southern Japan aboard a solid-propellant M-V-5 rocket on May 9, 2003, *Hayabusa* began a mission to the asteroid Itokawa, an Earth-approaching body named after Hideo Itokawa, the father of Japanese rocketry. Before liftoff, the spacecraft bore the name MUSES-C, an acronym of Mu Space Engineering Spacecraft ("Mu" designating the rocket, and "C" the third in its series). It weighed about 1169 lb (530 kg) at liftoff, and its box-like structure measured nearly 5 ft (1.5 m) on each side and almost 3 ft 6 in. (1 m) high.

Following a flyby of the Earth in 2004, *Hayabusa* arrived at the asteroid in September 2005, after a flight of 198 million miles (319 million km). But once it was there, not all went as expected. To begin with, *Hayabusa* hovered for some weeks over the asteroid, collecting data using Lidar, an X-ray spectrometer, and an infrared spectrometer. These observations would not only give context to the materials collected on the surface, but also helped in choosing a drop-off point for the lander, called *Minerva*. During the descent phase in November, the mission called for *Minerva* to touch down on Itokawa first, followed by *Hayabusa* itself. But the flight of *Minerva* proved to be problematical, as contact with it halted, and it probably floated off into space. An unsuccessful rehearsal with *Hayabusa* occurred later that month, during which the spacecraft landed, but a device that fired pellets to collect samples failed. *Hayabusa* returned again to gather bits of the asteroid, and it appeared at first that two sampling bullets (capable of collecting these tiny fragments) had been shot. But later telemetry raised questions as to whether this had really happened.

Then more problems materialized. Back in flight, *Hayabusa* went silent from early December 2005 to early March 2006. It seems that leaking thruster propellants affected the pointing of the spacecraft's antenna. Gradually, JAXA engineers and ground-controllers succeeded in establishing a signal and regaining mastery over the spacecraft, which had been spinning helplessly during the dark period. Fortunately, plenty of xenon remained to power the ion engine, and on April 25, 2007, controllers in Japan sent *Hayabusa* homeward, with a landing scheduled for 2010.

Exploring the Universe

The Search for Origins

4

Introduction

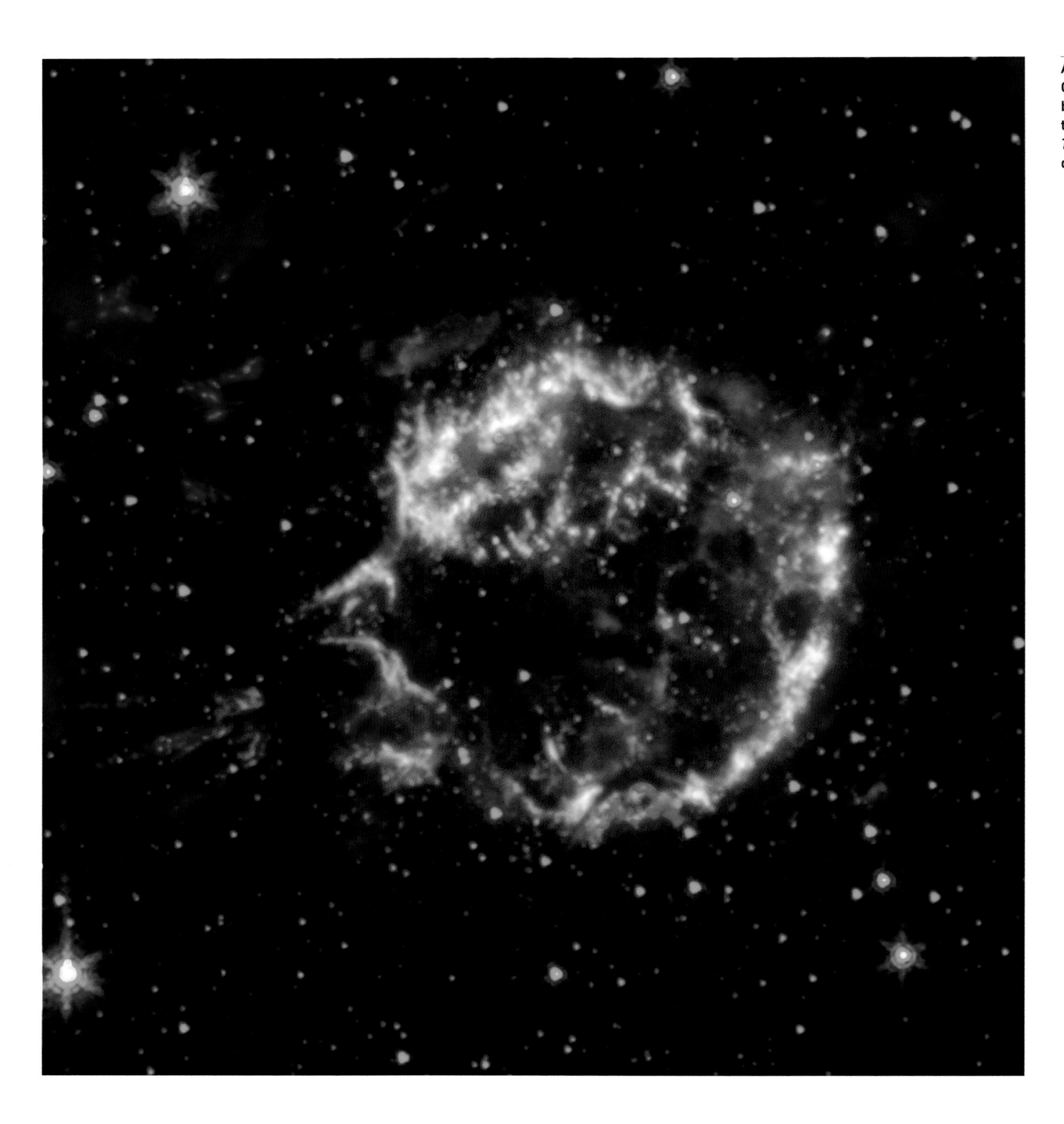

An exploded star named Cassiopeia A, observed by instruments aboard the *Spitzer Space Telescope*, scatters its debris in all directions.

Its eye open to the universe, the *Hubble Space Telescope* orbits at 353 miles (568 km) above the Earth. This photograph was taken just after its second servicing mission in 1997.

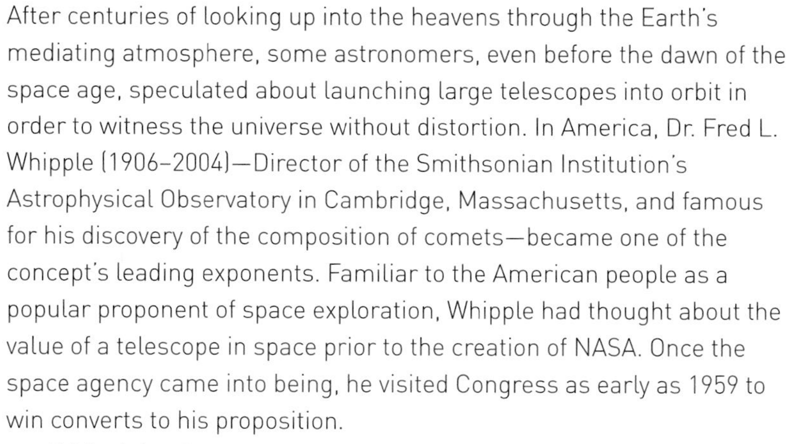

After centuries of looking up into the heavens through the Earth's mediating atmosphere, some astronomers, even before the dawn of the space age, speculated about launching large telescopes into orbit in order to witness the universe without distortion. In America, Dr. Fred L. Whipple (1906–2004)—Director of the Smithsonian Institution's Astrophysical Observatory in Cambridge, Massachusetts, and famous for his discovery of the composition of comets—became one of the concept's leading exponents. Familiar to the American people as a popular proponent of space exploration, Whipple had thought about the value of a telescope in space prior to the creation of NASA. Once the space agency came into being, he visited Congress as early as 1959 to win converts to his proposition.

Whipple's advocacy and that of others bore fruit in the realization of NASA's Orbiting Astronomical Observatory in 1968, in which Whipple had a generative hand. Despite mechanical failings, it gazed at the universe through ultraviolet, X-ray, gamma-ray, and infrared spectra. Three years earlier, the National Academy of Sciences' Space Science Board had lent its support, proposing that NASA dedicate a Saturn V rocket to lifting a large telescope into space. Elements of the National Academy of Sciences' recommendations appeared in the Skylab project (see p. 27). In it, NASA converted an upper stage of a Saturn V into an orbiting laboratory and installed an instrument called the Apollo Telescope Mount, a full-sized solar observatory trained permanently on the Sun. During the roughly eight months of Skylab's operating life (May 1973 to February 1974), astronauts maintained the telescope and targeted it on solar flares and other features.

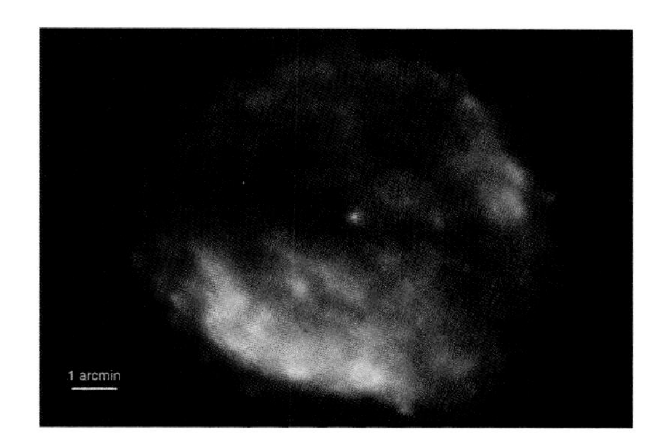

On August 23, 2005, *XMM-Newton* recorded this picture of a recent (2000-year-old) star explosion.

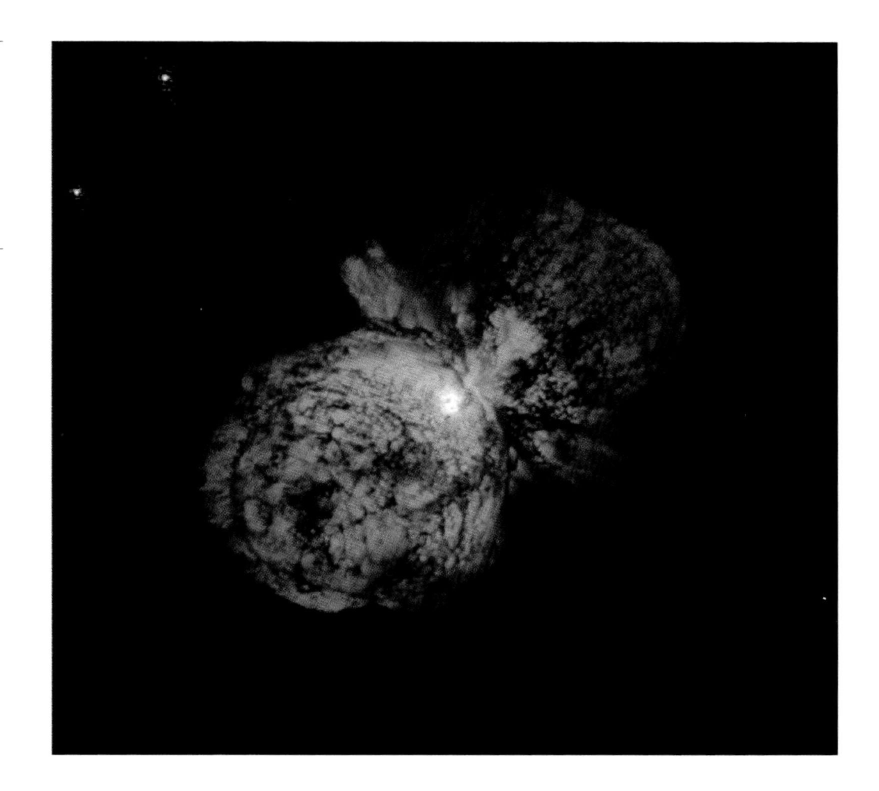

LEFT
The *Hubble Space Telescope* captured this image of a dust band swirling around the nucleus of the Black Eye Galaxy.

RIGHT
Hubble witnesses the death throes of the star Eta Carinae.

Hubble Space Telescope

Key facts

MANAGEMENT:	**NASA**
PRIME CONTRACTOR:	**Lockheed**
GO-AHEAD DECISION:	**1978**
LAUNCHED:	**April 1990**
FIRST REPAIR MISSION:	**December 1993**
ADDITIONAL MAINTENANCE / REPAIR / IMPROVEMENT MISSIONS:	**February 1997, December 1999, March 2002, September 2008 (scheduled)**
ESTIMATED CESSATION OF OPERATIONS:	**2013**
ORBIT:	**353 miles / 568 km above Earth**
LENGTH:	**43 ft / 13 m**
WEIGHT:	**24,000 lb / 10,890 kg**
INSTRUMENT:	**Telescope with 94-in. / 239-cm primary mirror**
PURPOSE:	**Make broad observations of the universe**

The confluence of two NASA projects that germinated during the 1970s transformed Fred Whipple's project from a mirage into an oasis. Without the Space Shuttle (see pp. 16–21), proponents of a big telescope had nothing to lift it into space. Without a big telescope project, the Space Shuttle backers had much less justification on Capitol Hill and elsewhere for their spaceplane. By mutually reinforcing each other, both projects came to life. But the two did not develop at the same pace. While the Shuttle program raced ahead during the 1970s, the big telescope project lay in wait. Only in 1978 came the formal go-ahead for an instrument that would have been beyond imagination just a few years earlier: 24,000 lb (10,890 kg) in weight, 43 ft (13 m) long, with a near-perfect 94-in. (239-cm) primary mirror. NASA named it after one of the world's great astronomers, Edwin Hubble (1889–1953), who, at the Mount Wilson Observatory outside of Los Angeles, had verified Albert Einstein's theory of an expanding universe, showed that the cosmos had not one but many galaxies, and developed a classification system for galaxies.

Despite its august name, the Hubble telescope underwent trial after trial. In 1983, the project's rising costs and delayed schedule forced NASA Administrator James Beggs to ask Congress for a bailout and an extension. With reluctance, the legislators acceded to both. However, the *Challenger* accident in January 1986 left mission managers no choice but to instruct the telescope's prime contractor, Lockheed, to put the almost completed instrument into storage until the Shuttle returned to flight. The *Hubble Space Telescope* (*HST*) finally went into orbit on April 24, 1990, aboard mission STS-31R from Kennedy Space Center's Launch Pad B. About twenty-four hours after liftoff, astronaut and astronomer Steven Hawley manipulated the

OPPOSITE
During the perilous first servicing mission to *Hubble* (December 1993), the Canadarm captured it and placed it upright in *Discovery*'s payload bay.

Technicians examine the *Hubble Space Telescope*'s 94-in. (239-cm) main mirror prior to launch.

Between December 4 and 8, 1993, four astronauts worked to restore the *HST* to its full potential during the initial servicing mission. This is a scene from this space drama.

***Hubble* floats against the backdrop of the Earth after its fourth repair mission (by the Shuttle *Columbia*), in 2002.**

OPPOSITE
The *HST* recorded a birthing scene in the Small Magellenic Cloud, where young stars are created.

Canadarm (see pp. 22–25) to hoist the mighty telescope—the size of a railway car—from the Shuttle's cargo bay and release it for an expected fifteen-year service life. But following the almost flawless deployment came weeks, months, and finally years of disappointment. Onboard computers started shutting down repeatedly, switching into safe mode as a result of the *HST*'s as yet unfamiliar motions. Then the unthinkable happened. The telescope seemed unable to focus on distant objects. Frantic inquiries revealed that the subcontractor who had ground the *Hubble*'s massive mirror had inadvertently fed incorrect specifications into the automated grinding machine, preventing the telescope from focusing all of the incoming light. Almost as bad, the spacecraft oscillated, confusing its pointing system (a problem, it later turned out, caused by the solar panels heating and cooling as they passed in and out of the Earth's shadow).

Caught in a hailstorm of adverse publicity, NASA acted. It planned for an unprecedented repair mission, a daring flight requiring many hazardous space walks with no assurance of success. The astronauts would be required not only to install corrective optics, but also to set up a suite of new computers, replace the existing wide-field planetary camera, take down the old and erect new solar arrays, and do sundry other complex tasks. Of course, all this had to be achieved while in orbit around the Earth. The chance came on December 2, 1993, when STS-61 carried the replacement modules and seven astronauts on a tense eleven-day mission. In all, during five long days between December 4 and 8, Story Musgrave, Jeffrey Hoffman, Kathy Thornton, and Thomas Akers used more than two hundred tools weighing an aggregate 14,400 lb (6530 kg) to accomplish the job.

Everyone knows the results (see p. 18). During the succeeding years, the *HST* astounded scientists and the public alike. Perched 353 miles (568 km) above the Earth, this unblinking giant has rewritten astronomical orthodoxy, perhaps more than any piece of equipment since Galileo trained his telescope on the skies some four hundred years ago and declared that the planets revolved around the Sun.

Since the first repair mission, the venerable *Hubble* has undergone renewal three more times. After the stunning success of 1993, astronauts returned again on STS-82 in February 1997, conducting maintenance and expanding the telescope's range into near infrared. In December 1999, STS-103 voyaged to *Hubble*, ostensibly on a mission of preventive maintenance, but once there the crew found themselves replacing all six gyroscopes (three had failed) and installing a new main computer. The fourth servicing mission occurred in March 2002, with the STS-109 team putting into operation the Advanced Camera for Surveys, which is capable of seeing from visible to far ultraviolet wavelengths. Then, in October 2006, NASA Administrator Michael Griffin approved a last flight to *Hubble*, scheduled for 2008, to extend its operations to 2013 (by which date the instrument will have been in service for twenty-three years). The visitors will install a new Cosmic Origins Spectrograph and the Wide Field Camera 3, and replace gyroscopes and batteries. Meanwhile, NASA engineers and scientists planned for the launch, also in 2013, of the James Webb Space Telescope, named for the NASA leader who guided the Apollo program. Mainly an infrared-range machine, it will be placed in orbit about 1 million miles (1.6 million km) from Earth.

TOP
The *HST* reveals (in infrared) the so-called Sombrero Galaxy.

ABOVE, CENTER
***Hubble* captures the soft, pearly radiance of Galaxy M81.**

ABOVE
Improved by the new main computer installed during the third servicing mission in 1999, *Hubble* recorded this image from the heart of the Whirlpool Galaxy.

RIGHT
A towering stellar spire in the Eagle Nebula, visible to the *HST* during 2005.

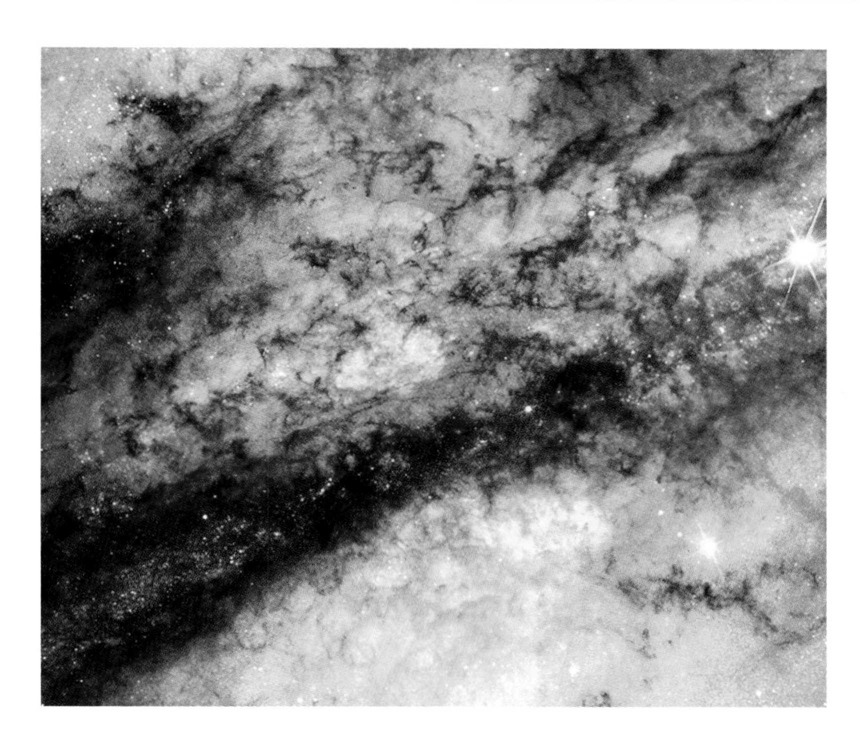

LEFT
The *HST* peers into the nucleus of Galaxy Centaurus A.

RIGHT
Brilliant as it died, the Cat's Eye Nebula was caught by *Hubble* exploding in a kaleidoscope of gas and dust.

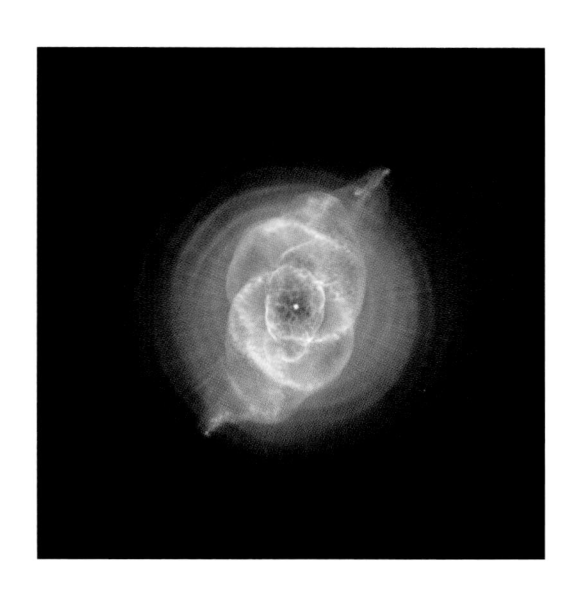

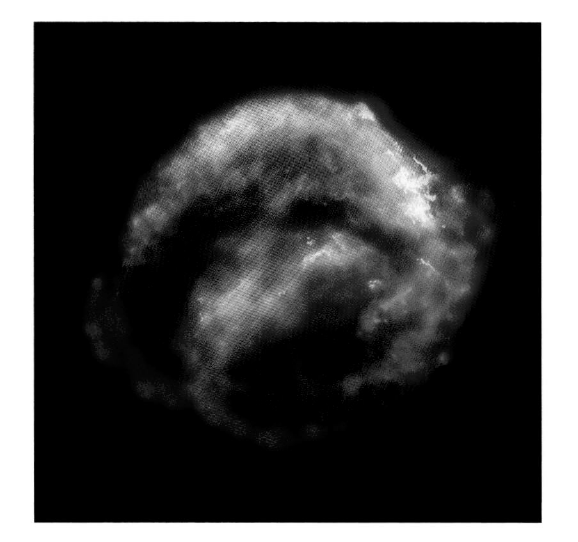

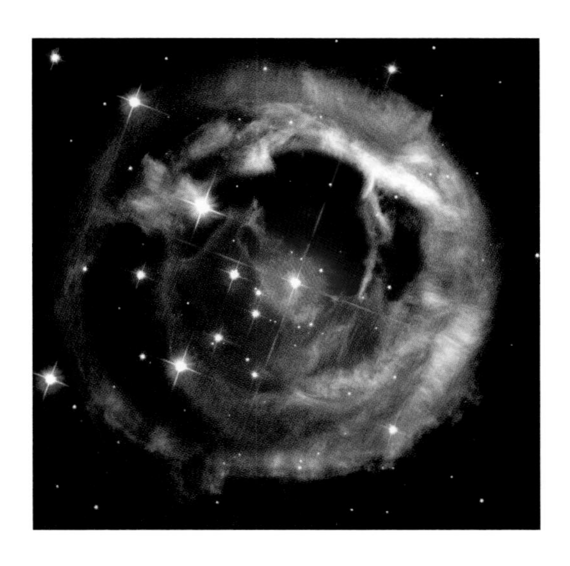

ABOVE, CENTER
Fitted on the fourth servicing mission with a camera that captures images from visible to far ultraviolet, *Hubble* subsequently took this picture of Kepler's Supernova Remnant.

ABOVE
In December 2002, the *HST* gathered light radiated from the red supergiant star V838 Monocerotis.

Proof of repair: this panoramic image of the center of the Orion Nebula was created by the *Hubble* team during the year after the first servicing mission to the telescope.

Ulysses

An artist's rendering of the spacecraft *Ulysses* passing through the tail of Comet Hyakutake in 1996.

Key facts

MANAGEMENT:	**ESA and NASA**
LAUNCHED:	**October 1990**
PASSED OVER THE SUN'S POLES:	**1994, 1995, 2000, 2001, 2007, 2008 (scheduled)**
OBSERVED COMET HYAKUTAKE:	**May 1996**
DIMENSIONS:	**10 ft / 3 m long, 11 ft / 3.4 m wide, 7 ft / 2.1 m high**
WEIGHT:	**808 lb / 366 kg**
PURPOSE:	**Make polar observations of the Sun's behavior**

NASA and the European Space Agency (ESA) joined forces to launch a spacecraft dedicated to unraveling the secrets of our star, the Sun, and by inference those of other stars as well. Like its Greek namesake, who spent ten years traveling through unfamiliar regions after the fall of ancient Troy, *Ulysses* set forth on a great journey, to chart the unknown portions of space above and below the poles of the Sun. The spacecraft itself and half of the instruments on board originated from the ESA countries; NASA supplied the other half, as well as launch support from the Space Shuttle *Discovery* (STS-41), and commands the project through the Deep Space Network run by the Jet Propulsion Laboratory (JPL).

Once released from *Discovery*'s cargo bay on October 6, 1990, the spacecraft pointed toward Jupiter and flew to its target, boosted by an Inertial Upper Stage (IUS), and later by a Payload Assisted Module (PAM). It advanced on the planet with the highest velocity of any object fashioned by humans (until *New Horizons*, see pp.122–23), separating from the PAM once in interplanetary orbit. *Ulysses* reached Jupiter in February 1992, and, after making observations of the planet's magnetosphere during the flyby, used gravity-assist to swing into polar orbit of the Sun. The spacecraft passed over the Sun's poles in 1994 and 1995, ending its primary mission in September 1995. But ESA and NASA continued the voyage of *Ulysses*. It observed the ion tail of Comet Hyakutake in May 1996, returned to the Sun's poles in 2000 and 2001, and after more years of travel visited again in 2007 and early 2008. But the two space agencies decided in February 2008 to end Ulysses' mission after seventeen years of service because of the effects of the extreme solar climate.

A squat box measuring roughly 10 ft (3 m) long, 11 ft (3.4 m) wide, and 7 ft (2.1 m) tall, *Ulysses* weighed a petite 808 lb (366 kg) at launch (instruments accounting for just 121 lb/55 kg of the total weight). Its mission planners hoped the spacecraft's long sojourn would illuminate such mysteries as the origins of the solar wind, including its accompanying shocks and waves; the genesis of, and the processes influencing, cosmic rays; the motions of energetic particles; the nature of interstellar and interplanetary dust; the sources of gamma-ray bursts; and the three-dimensional attributes of the magnetic field as related to the solar wind. All of these factors vary with the eleven-year solar activity cycle, and part of *Ulysses*' mission involves pinpointing how the Sun's states of quiescence and activity impact those parts of space under its influence.

To do so, *Ulysses* had on board an array of U.S. and European instrument suites. The Americans' equipment observes solar wind plasma, low-energy ions and electrons, cosmic rays and solar particles, solar x-rays and gamma-ray bursts, and radio and plasma waves. ESA's researchers concentrated on the magnetic field, energetic particles and interstellar neutral gas, cosmic dust, radio science coronal sounding, gravitational waves, directional discontinuities, and mass loss and ion composition. Solar wind composition measurements involved both U.S. and European researchers.

By 2007, many scientific inferences had been drawn, based on the data streaming home from *Ulysses*. But much remained to be learned. For instance, during the third overflight of the Sun's South Pole, in February 2007, researchers discovered to their surprise that a disparity in polar temperatures found during the 1994–95 mission had persisted. Evidently, as measured in Fahrenheit, the North Pole remained about 80,000 degrees cooler than the South—a difference of roughly 8 percent. In addition, *Ulysses* may help scientists address other curiosities, such as one involving the Earth's magnetic poles. On the Sun, the magnetic poles reverse direction (North and South change places) every eleven years, corresponding to sunspot activity. The same phenomenon occurs on Earth every 300,000 years or so, and although the reasons remain unclear, *Ulysses*' data may offer insights. Finally, *Ulysses* made tantalizing observations of such unexpected and unexplained features as coronal holes over the Sun's poles, mysterious zones where the star's magnetic field opens, and galactic cosmic rays pour in as solar winds rush out.

Ulysses approaches the Sun in this artist's concept. The year 2008 will be the spacecraft's eighteenth year of space travel.

LEFT
Technicians at ESA's testing facilities put the _Ulysses_ spacecraft through its paces.

RIGHT
Ulysses undergoes tests at the European Space Research and Technology Center, Noordwijk, The Netherlands.

Solar and Heliospheric Observatory

A portrait of the *SOHO* spacecraft, standing 14 ft (4.3 m) tall, during its assembly phase.

Key facts

MANAGEMENT:	**ESA and NASA**
LAUNCHED:	**December 1995**
ORBIT:	**Synchronous with the Earth, 945,000 miles / 1.5 million km distant**
WEIGHT:	**4079 lb / 1850 kg**
DIMENSIONS:	**14 ft / 4.3 m long, 12 ft / 3.7 m wide, 9 ft / 2.7 m deep**
PURPOSE:	**Orbit a "stationary" platform to observe the Sun's interior, atmosphere, and wind over long periods**
ESTIMATED CESSATION OF OPERATIONS:	**December 2009**

Ulysses (see pp. 150–51) was not an isolated attempt to comprehend solar phenomena; it was the first of a constellation of spacecraft aimed at grasping the workings of our neighboring star. Five new projects will enter the solar sweepstakes, one from Japan (*Solar B*), one from Europe (*Proba 2*), and three from NASA (the STEREO twins and the *Solar Dynamics Orbiter*). Some years from now—perhaps in 2015—the global International Living with Star project may launch a solar orbiter to circle the Sun from close range.

As planning for these projects got underway, *Ulysses* continued its pioneering work, and was accompanied five years after its initiation by a second European probe called the *Solar and Heliospheric Observatory*, or *SOHO*. On December 2, 1995, an Atlas II rocket lifted off from Cape Canaveral Air Force Station, Florida; two hours after launch, the rocket's Centaur upper stage was fired. This stack pushed *SOHO* toward its mission as a "stationary" platform in space, from which scientists could observe the outer and inner workings of the Sun for at least two years. For *SOHO*, the Europeans took the role of senior partner. ESA industries developed the spacecraft and nine of its twelve instruments. Aside from launch services, NASA contributed communications and daily operations from Goddard Space Flight Center, and furnished three instruments. The spacecraft itself consists of two pieces. The bottom part, called the Service Module, holds "housekeeping" equipment for such operations as power generation, thermal control, pointing and telecommunications, and solar panel support. Atop it, the Payload Module contains the scientific instruments. Relatively big and heavy, *SOHO* measures 14 ft (4.3 m) tall, 9 ft (2.7 m) deep, and 12 ft (3.7 m) wide. It weighed 4079 lb (1850 kg) at launch, only 1345 lb (610 kg) of which constituted the payload.

Perhaps the distinguishing feature of *SOHO*—apart from its mission—rests with the mechanics of its positioning. In effect, the spacecraft provides an unblinking view of the Sun, around which it revolves in synchronization with the Earth, but at a distance of about 945,000 miles (1.5 million km) from our planet. *SOHO* actually rotates in a six-month orbit around the so-called first Lagrange Point, where the combined gravity of the Sun and Earth lock it in an Earth–Sun line, enabling it to orbit the Sun in the same period of time as the Earth. *SOHO* represents the first space observatory to view the Sun from a point other than Earth orbit.

The *SOHO* team (a patchwork of twenty-nine institutes from fifteen countries) concentrates on three avenues of research: the Sun's interior, its atmosphere, and its wind. To see inside the star, *SOHO* uses instruments that measure oscillations inside it, even down to its nucleus, including GOLF (Global Oscillations at Low Frequencies), manufactured in France; VIRGO (Variability of Solar Irradiance and Gravity Oscillations), from The Netherlands; and MDI/SOI (Michelson Doppler Imager/Solar Oscillations Investigation) from the U.S. (for the outer layer of the Sun's interior). Five of *SOHO*'s instruments watch the Sun's atmosphere: SUMER (Solar Ultraviolet Measurements of Emitted Radiation) from Germany; CDS (Coronal Diagnostics Spectrometer) from the U.K.; EIT (Extreme Ultraviolet Imaging Telescope) from the U.S.; UVCS (Ultraviolet Coronagraph Spectrometer) from the U.S.; and LASCO (Large Angle and Spectrometric Coronagraph experiment) from the U.S. Finally, the solar wind is detected by CELIAS (Charge, Element, and Isotope Analysis System) from Switzerland; COSTEP (Comprehensive Suprathermal and Energetic Particle Analyzer) from Germany; ERNE (Energetic and Relativistic Nuclei and Electron experiment) from Finland; and SWAN (Solar Wind Anisotropies) from Finland.

At the ten-year mark, *SOHO* continued to function well. However, the team did experience one tense period from mid-1998 to early 1999. First, they lost control of *SOHO*. They recovered it with no serious damage, but then two of its three gyroscopes failed immediately, and one later on. In response, they uploaded new software, enabling the spacecraft to return to full operation, and making it the only three-axis stabilized spacecraft flying with no gyroscopes. The ESA Science Program Committee decided in May 2006 to extend *SOHO*'s life from the previous end date of April 2007 to December 2009, at which point it will have been in operation for more than fourteen years. In the meantime, it provides scientists with vast amounts of data relating to such areas of investigation as the structure of sunspots below the surface; the precise measurement of solar temperatures; the acceleration of fast and slow solar winds; the identification of hitherto unknown phenomena, such as coronal waves and solar tornadoes; the early warning of solar occurrences having direct influences on activities on Earth; and the monitoring of the total radiation emanating from the Sun. *SOHO* has also been instrumental in the discovery of more than one thousand comets.

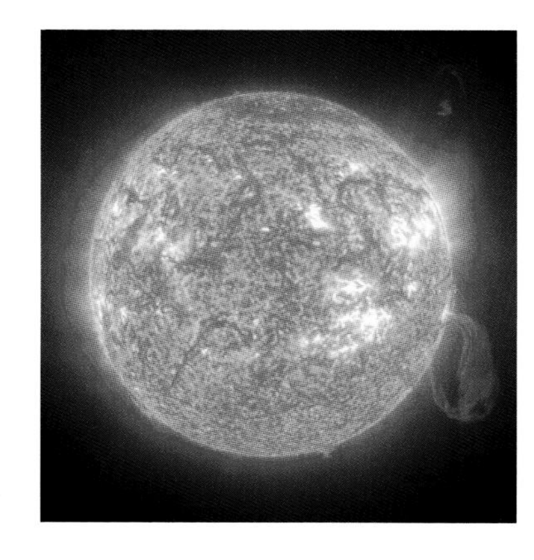

With its eye fixed on the Sun, *SOHO* collects images of solar activity, such as this bulge on the Sun's profile, indicating huge clouds composed of relatively cool, dense plasma.

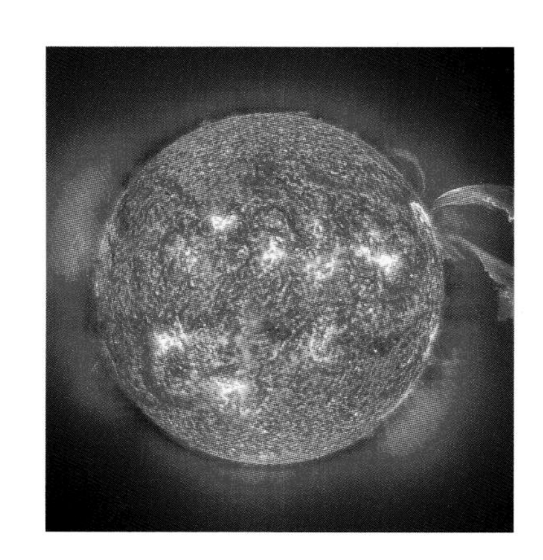

The *SOHO* spacecraft reveals the cauldron that is our Sun, here recording two curved tongues of escaping plasma.

Far Ultraviolet Spectroscopic Explorer

***FUSE* awaits launch from Cape Canaveral Air Force Station, Florida.**

In order to accomplish its mission to explore the universe, NASA has engaged in many different types of partnerships. *SOHO* (see pp. 152–53) involved the U.S. space agency in launch, ground-control operations, and a share of the onboard experiments. NASA contributed to *Ulysses* (see pp. 150–51) by providing the booster, use of the JPL's Deep Space Network, and about half of the experiments on board the spacecraft itself. In both cases, NASA and ESA negotiated their respective roles and then achieved the collaboration. By contrast, conceiving, building, and operating the *Far Ultraviolet Spectroscopic Explorer* observatory (*FUSE*) involved a complicated organizational structure with many participants. In part, this situation resulted from the "faster, better, cheaper" initiative pursued by former NASA Administrator Daniel Goldin during the 1990s, in which the funding of many projects was reduced, leaving project managers to cobble together whatever alliances they could.

A team of Johns Hopkins University scientists developed and operated *FUSE* in cooperation with colleagues from the University of Colorado at Boulder and the University of California, Berkeley. In addition, three governmental partners—NASA, the French Space Agency (CNES), and the Canadian Space Agency (CSA)—contributed, as did corporate entities such as Allied Signal and the Orbital Sciences Corporation. Johns Hopkins itself provided project management, as well as satellite and science planning and operations. Berkeley pursued far ultraviolet detector development, Colorado far ultraviolet spectrograph development, and the two university teams integrated their projects into the spacecraft as a whole. CNES provided holographic gratings for *FUSE*, the CSA pursued the Fine Error Sensor packages, and Goddard Space Flight Center managed the guest investigator program and contract management. Orbital Sciences Corporation fabricated the spacecraft and Allied Signal built the ground station and handled on-orbit operations. Finally, NASA's Headquarters Office of Space Science Applications funded the $108 million project (reduced from an original estimate of $350 million) through its Origins program. *FUSE* represents the biggest astrophysics project NASA has ever delegated to a university.

FUSE entered Earth orbit on June 24, 1999, atop a Delta II launch vehicle, initially for three years of operation. It flies 475 miles (764 km) above our planet in a circular path. Within its towering 18-ft (5.5-m) structure (which weighs 3000 lb/1360 kg), it houses two sections: the spacecraft and the mission's primary instrument. The spacecraft portion holds everything necessary for the satellite to function: power, the pointing mechanism, attitude controls, the solar panels, electronics, antennas, and communications equipment. The instrument

Key facts

MANAGEMENT:	**Johns Hopkins University, University of Colorado at Boulder, University of California Berkeley, NASA, CNES, and the CSA**
MANUFACTURE:	**Orbital Sciences Corporation**
ON-ORBIT OPERATIONS:	**Allied Signal Corporation**
LENGTH:	**18 ft / 5.5 m**
WEIGHT:	**3000 lb / 1360 kg**
LAUNCHED:	**June 1999**
PRIMARY INSTRUMENT:	**Four-mirror telescope**
PURPOSE:	**Observe far ultraviolet explosions in the universe related to the "Big Bang"**

consists of four telescope mirrors that are specially coated with silicon carbide and lithium fluoride and aligned together, gathering light from four optical channels in the far ultraviolet range. The equipment operates at high resolution, enabling researchers to isolate and spread out key features in the spectrum. Most other telescopes do not function in this very specific region of light, and the last one that did (the *Copernicus* satellite during the 1970s) operated at 10,000 times less sensitivity than *FUSE*.

FUSE exists to shed light on the formative period of the universe. The project has succeeded in collecting important data relating to the so-called "Big Bang," asking what conditions may have been like a few minutes afterward, how the consequent chemical elements have been dispersed across the galaxies, and how these elements influence the evolution of galaxies. In particular, the *FUSE* researchers are hunting for deuterium, a heavy hydrogen formed exclusively during the Big Bang, as well as the hot gas content of our own Milky Way and that of its nearest neighbor, the Magellanic Clouds.

FUSE has not been free of malfunctions. Late in 2001, and again in late 2004, reaction wheels in the satellite's pointing mechanism failed. Software workarounds saved the day both times. In spring 2007, one of the reaction wheels stopped spinning yet again, but this time resumed after a month's hiatus. Despite these difficulties, after the three-and-a-half-year primary mission ended on April 1, 2003, *FUSE* went into an extended period of operation and collected masses of data on the foundations of the universe, before being shut down in October 2007.

An artist's conception of *FUSE* in space, hunting for traces of such chemicals as deuterium, present at the time of the universe's creation.

PAGES 156–57
A *FUSE* image of the star AE Aurigae (the bright point of light at the center of the picture), embedded in a portion of space containing carbon-saturated dust grains.

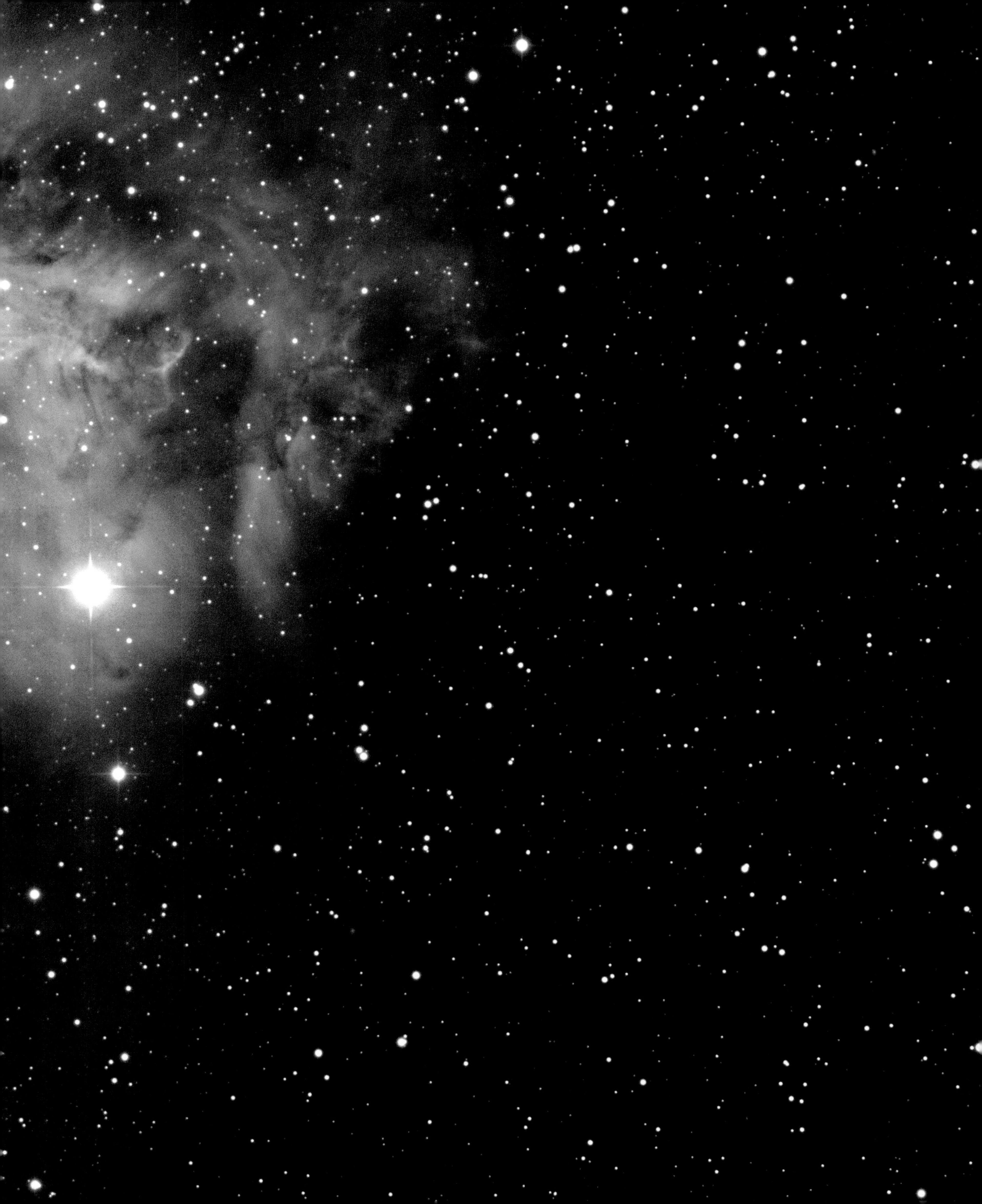

Chandra X-ray Observatory

Key facts

FIRST PROPOSED:	**1976**
MANAGEMENT:	**NASA Marshall Space Flight Center**
PRIME CONTRACTOR:	**TRW**
LAUNCHED:	**July 1999**
WEIGHT:	**12,930 lb / 5865 kg**
LENGTH (WITH SUNSHADE):	**45 ft / 13.7 m**
PRIMARY INSTRUMENT:	**Telescope with eight tubular mirrors**
PURPOSE:	**Make X-ray observations of black holes and other phenomena**

The *Chandra X-ray Observatory*—like such other costly and famous projects as the *Hubble Space Telescope* (see pp. 144–49)—had a long and unpredictable genesis. It evolved over twenty-three years, starting in 1976 with a proposal to NASA by Dr. Harvey Tananbaum of Harvard University and Dr. Riccardo Giacconi of the Johns Hopkins University. The following year, NASA's Marshall Space Flight Center and the Smithsonian Astrophysical Observatory joined forces to bring it to fruition. Originally, *Chandra* had been conceived as a project parallel to the *HST* in which, poised in Low Earth Orbit, the big machine would be serviced and maintained by occasional Shuttle missions. The project progressed through the early 1980s, during which time the mirror and technologies underwent developmental research. Fabrication began later in the decade, but in 1992 budgetary pressures reduced *Chandra*'s scope. After a redesign, what had been twelve mirrors became eight, and the anticipated six scientific instruments became four. In addition, mission planners decided to launch *Chandra* into a high elliptical orbit—a third of the distance to the Moon (87,000 miles/140,000 km) at its farthest point—beyond the prospect of repair by the Shuttle, but also safely above the Earth's radiation belts. During the mid-1990s, *Chandra*'s mirrors underwent much travel: to Connecticut for polishing by Raytheon, to California to be coated by the Optical Coating Laboratory, and to New York for assembly by Eastman Kodak. Next, NASA's Marshall Space Flight Center engineers tested and calibrated the mirrors as well as *Chandra*'s science instruments. By March 1998, TRW had completed the overall assembly of the observatory.

OPPOSITE
***Chandra* in orbit aboard *Columbia* in July 1999, just before being inclined upward from the cargo hold and released.**

RIGHT
At Launch Pad 39A at the Kennedy Space Center, technicians prepare to close the *Chandra* telescope behind the payload-bay doors of the Shuttle *Columbia*, shown here prior to liftoff.

FAR RIGHT
Held fast in the Vertical Processing Facility at the Kennedy Space Center, the *Chandra X-ray Observatory* is about to be mounted on the inertial upper stage booster, which is beneath it.

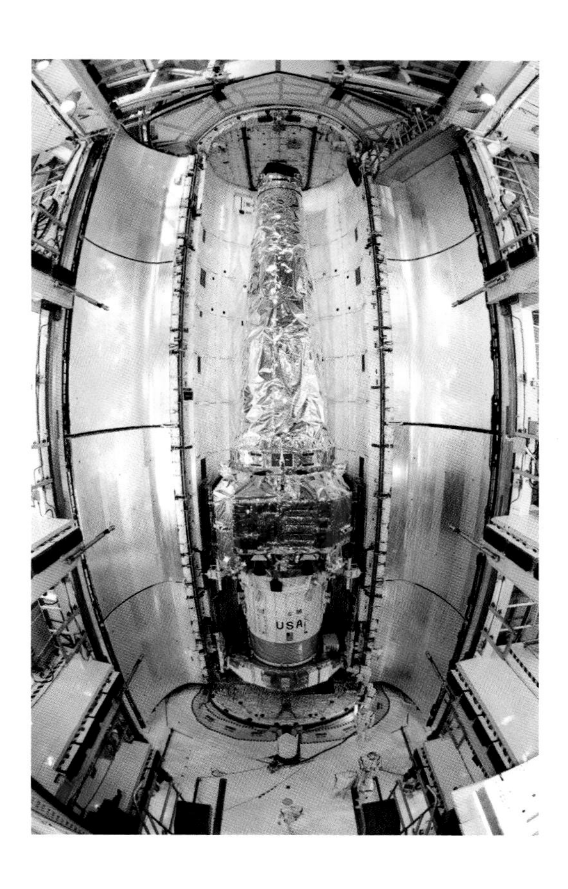

The *Chandra* and *Hubble* telescopes joined forces to witness a fiery dead star in the Crab Nebula.

RIGHT

In the Perseus arm of the Milky Way, some 6000 light years from the Earth, *Chandra* observed many immense stars forming in a string of stellar clusters in the W3 region.

FAR RIGHT

In cooperation with the *HST*, Chandra penetrated the dark and ethereal Pillars of Creation, inside the Eagle Nebula (M16).

The name of the telescope commemorates the Indian-born astrophysicist Dr. Subrahmanyan Chandrasekhar (1910–1995), who studied at the University of Madras, India, and at Cambridge University, England. Chandrasekhar taught at the University of Chicago, beginning in 1938, and became an American citizen in 1953. His research concentrated on supernovas, neutron stars, and black holes, and culminated in 1983 in the Nobel Prize for Physics.

Chandra lifted off from Kennedy Space Center on July 23, 1999, aboard STS-93. It represented the heaviest and largest object boosted by any Shuttle flight (at least until mid-2006). The instrument itself weighed 12,930 lb (5865 kg) at launch, but the IUS added another 32,500 lb (14,740 kg). A two-stage solid-fuel booster, the IUS would propel *Chandra* from Earth orbit to the more distant loop desired by mission planners. Not only did this payload rank alongside the heaviest, it had big dimensions as well. The IUS by itself measured 17 ft (5.2 m) in length. The *Chandra* was a whopping 45 ft (13.7 m) long with sunshade open, and 64 ft (19.5 m) wide with its solar arrays deployed. The full length of the mated IUS–*Chandra* stack was 57 ft (17.4 m), and the total weight (with all support equipment) 50,162 lb (22,753 kg).

The roughly cone-shaped *Chandra* observatory consists of three parts: the spacecraft, the telescope, and the science instrument module. The spacecraft portion, located in a collar near the front of the observatory, contains computers and communications systems that allow transmissions between ground stations and *Chandra*, and constitutes the command and control mechanism of the whole machine. This compartment also holds a camera that orients the telescope, machinery that operates the solar panels, and propulsion rockets to move and aim the observatory. The telescope itself represents a long leap in capability, with eight times greater resolution and twenty to fifty times greater sensitivity than previous X-ray models. This advantage results in part from its set of eight tubular mirrors, the biggest and smoothest ever produced, the largest being 3 ft (0.9 m) long and 4 ft (1.2 m) in diameter. The entire mirror group weighs more than 2000 lb (907 kg). Moreover, the reflective covering over the entire exterior of the telescope has been integrated with heating units inside the instrument to provide a constant temperature and more accurate observations. *Chandra*'s science instruments represent the third part of the observatory. The High Resolution Camera, at the rear of the observatory, records X-ray images, specifically of high-energy events like the death of stars. The Advanced X-Ray Astrophysics Facility (AXAF) Charge Coupled Device (CCD) Imaging Spectrometer records the color, or energy, of the X-rays.

Chandra operates on a complex web of partnerships. The Marshall Space Flight Center manages the project for NASA's Headquarters Science Mission Directorate. The Smithsonian Astrophysical Observatory plans the science missions, based upon proposals from the scientific community. The Chandra X-Ray Center—a consortium involving the Smithsonian Astrophysical Lab, the Massachusetts Institute of Technology (MIT), and TRW (the *Chandra* prime contractor)—operates the Chandra Operations and Control Center. The JPL provides communications via its Deep Space Network.

Chandra's initial period of flight ended in 2004, five years after its launch, but it continues to train its eyes on the universe. *Chandra* examines distant galaxies that reverberate with explosions, and which perhaps contain vast black holes at their centers, giving scientists insights into their intense bursts of energy. It also offers clues about dark matter, a mysterious substance without which the hot, X-ray-emitting gas between the galaxies and the clusters themselves would lack sufficient mass and gravity to cohere. *Chandra* has made historic observations, such as the first identification of a binary black hole, the first sound waves detected from a black hole, the deepest X-ray exposure ever made, a possible black hole in the Milky Way, and the most distant X-ray cluster ever detected.

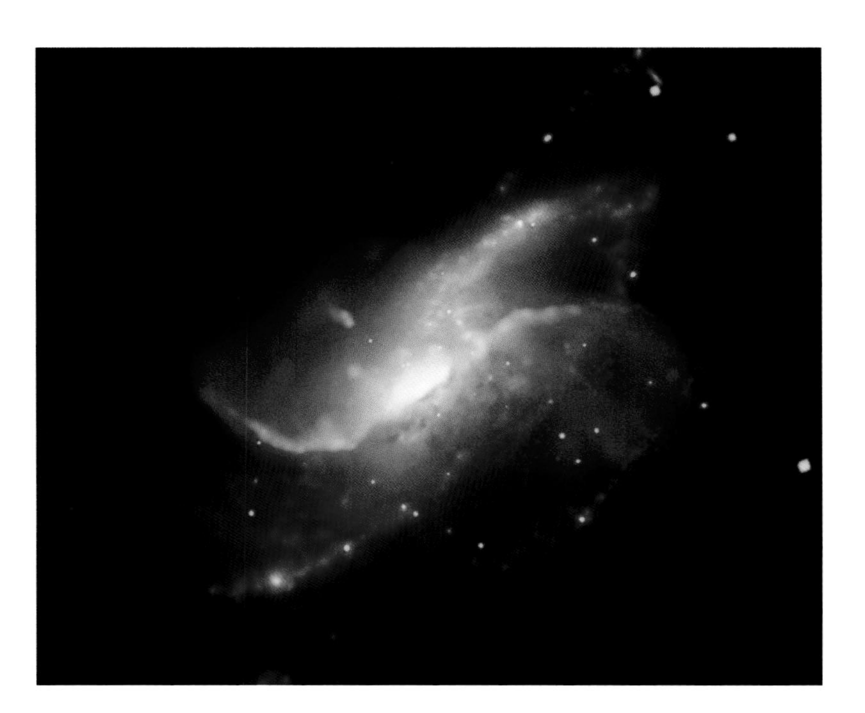

LEFT
***Chandra*'s X-ray vision and the *Spitzer Space Telescope*'s infrared vision combined to form this image of Spiral Galaxy M106.**

RIGHT
The *Chandra X-ray Observatory* witnesses the disintegration of a star: the hot gas in the middle of 3C442A (depicted in blue) pushes the two ends apart.

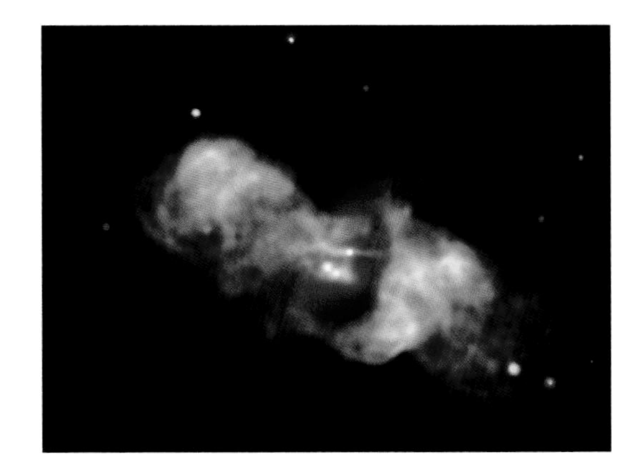

XMM–Newton

An artist imagines the *XMM–Newton* spacecraft in flight, showing its impressive dimensions.

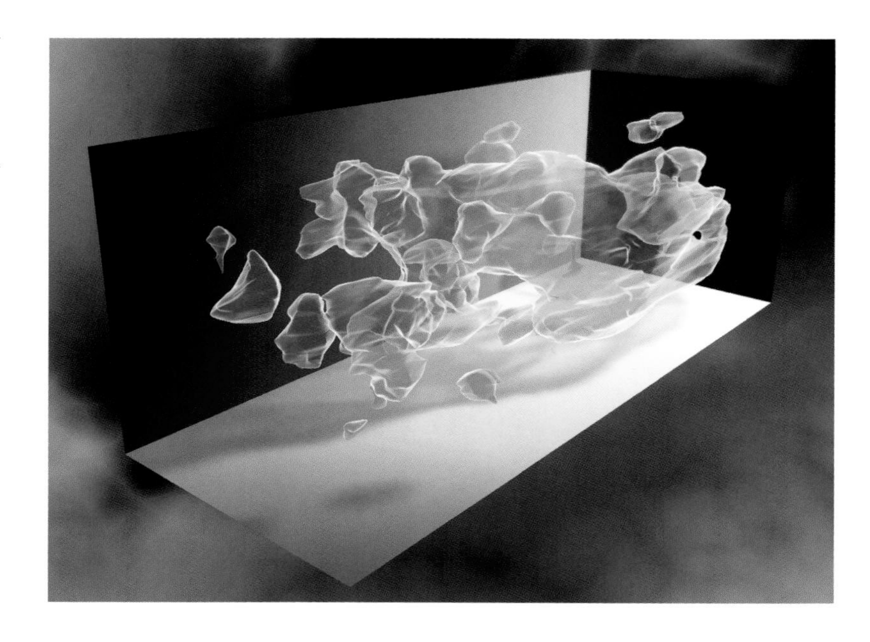

XMM–Newton embodies the disembodied, giving substance to dark matter in this three-dimensional distribution.

Key facts

MANAGEMENT:	ESA
LAUNCH:	December 1999
PURPOSE:	Make X-ray observations of black holes and other phenomena
PRIMARY INSTRUMENTS:	Three telescopes with 170 mirrors (collectively)
LENGTH:	32 ft / 9.8 m
WEIGHT:	8379 lb / 3800 kg
EXPECTED CESSATION OF OPERATIONS:	2010

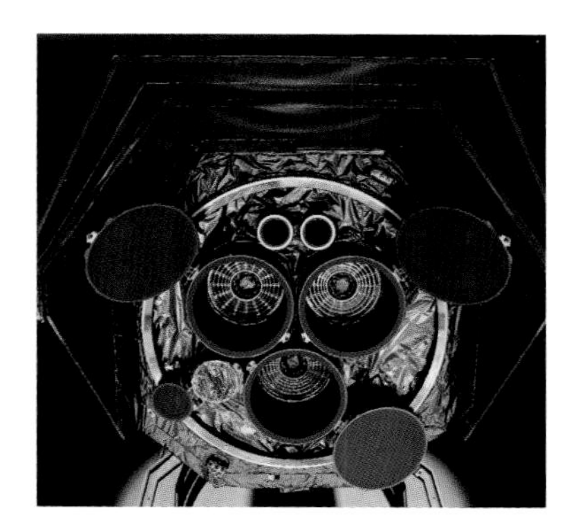

A view of ESA's *XMM-Newton* spacecraft during assembly, showing its X-ray telescopes.

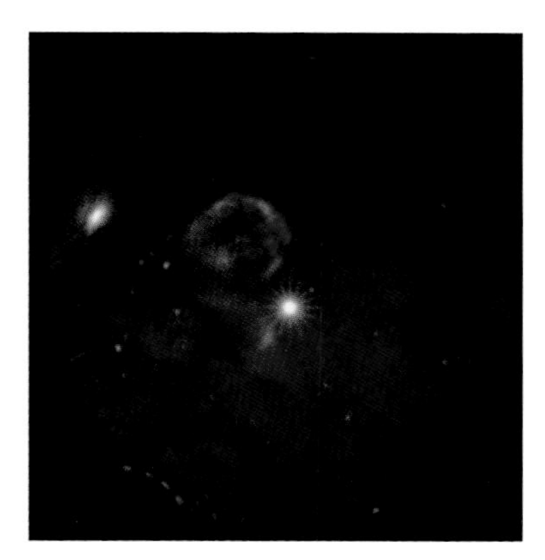

LEFT
***XMM-Newton*'s three X-ray imaging cameras took this composite color picture of the Supernova SN 1987A.**

RIGHT
This ultraviolet image of M81 was captured by *XMM-Newton*'s Optical Monitoring Telescope.

Not to be outdone by NASA and its partners, on December 10, 1999—just five months after *Chandra* (see pp. 158–61) went into orbit—ESA launched the *XMM-Newton* satellite from Kourou, French Guiana, aboard an Ariane 5 (the first commercial flight of this rocket). As with *Chandra*, *XMM-Newton*'s builders heralded it as an X-ray satellite observatory like no other, stating that it was "the biggest scientific satellite ever built in Europe. Its telescope mirrors are the most sensitive ever developed in the world, and with its sensitive detectors, it sees much more than *any previous X-ray satellite*." The ESA team also claimed that "the highly nested X-ray Multi-Mirrors . . . are enabling astronomers to discover more X-ray sources than any of the previous space observatories."[1] Even its name—honoring the towering figure of physics and mathematics, Sir Isaac Newton (1642–1727)—seemed to throw down the gauntlet.

The other part of the satellite's name—XMM, for X-Ray Multi-Mirror Design—denoted a significant technological difference from *Chandra*. *XMM-Newton* transported not one but three advanced, barrel-shaped telescopes, and not eight but more than one hundred and seventy precision, cylindrical mirrors spread out over the three telescopes (each mirror roughly 12 to 28 in./30 to 71 cm in diameter, and 24 in./61 cm long), all carefully arrayed one inside another, like Russian nesting dolls, to offer the maximum collecting area for the elusive X-rays. Moreover, the mirrors represent a technical breakthrough, not only because they are wafer-thin, but also (according to project officials) because they are the smoothest of any type yet created. The spacecraft also came equipped with three X-ray imaging cameras (each one focused in synchronization with one of the three telescopes), two spectrometers, and an optical monitoring telescope aligned with the main X-ray telescope for simultaneous observations in the visible and X-ray bands. In fact, all of the spacecraft's instruments operate in parallel, a technique learned from ESA's 1980s-era X-ray satellite, *Exosat*. *XMM-Newton*'s designers had sufficient confidence in its durability to predict an operational life of up to ten tears. Indeed, six years into its service life, ESA extended *XMM-Newton*'s mission to March 2010.

Like *Chandra*, *XMM-Newton* had a long gestation period. Conceptual work on the satellite began in 1982, and two years later the ESA states committed themselves to 50 percent bigger budgets for four "Cornerstone" projects, *XMM-Newton* being one. The project started formally in 1985 and the design phase ended in 1987. ESA approved the mission in June 1998. In all, fourteen countries and forty-six companies contributed to *XMM-Newton*. The satellite also has stately proportions. It is 32 ft (9.8 m) long, more than 52 ft (15.9 m) wide with solar panels open, and weighs 8379 lb (3800 kg). Of course, grand as they may be, these dimensions do not measure up to *Chandra*'s (which, in the context of launch costs, may actually be high praise).

Also like *Chandra*, *XMM-Newton* casts its gaze on the fiery parts of the universe, concentrating on, among others, neutron stars, whose extreme density and temperatures may resemble conditions existing a fraction of a second after the Big Bang, when the essential ingredients of life went flying in all directions. As a result of the satellite's observations, scientists are measuring for the first time the impact of a neutron star's extremely powerful gravitational field on its own light. These findings have revealed not only the strength of this gravitational pull, but also the degree of compaction of these marvels of the universe.

Suzaku

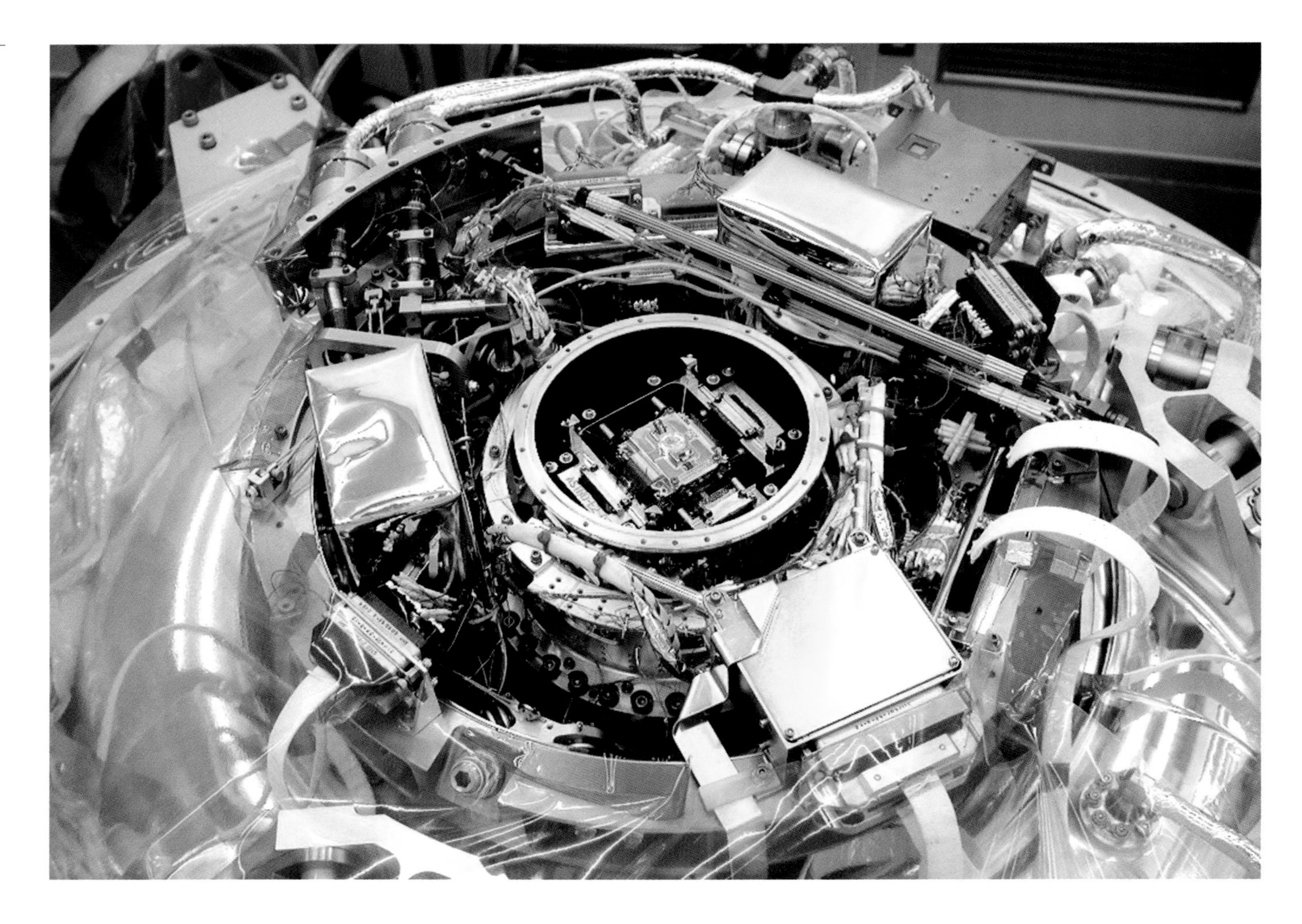

The front-end assembly of JAXA's *Suzaku* spacecraft.

A photograph of the fully assembled *Suzaku* spacecraft prior to undergoing vibration tests.

Key facts

LAUNCHED:	**July 2005**
MANAGEMENT:	**JAXA**
ORBIT:	**345 miles / 555 km above Earth**
LENGTH:	**23 ft / 7 m**
WEIGHT:	**3500 / 1590 kg**
PURPOSE:	**Make X-ray observations of black holes and other phenomena**
PRIMARY INSTRUMENTS:	**Five X-ray telescopes**

The Japanese Aerospace Exploration Agency (JAXA) fielded its own candidate in X-ray science, known, after it went into orbit on July 10, 2005, as *Suzaku* (in Japanese, the "red bird of the south," and in Chinese mythology, a sparrow-like bird that wards off evil and brings good fortune). Prior to launch, the satellite bore the name Astro-E2, successor to Astro-E, whose booster failed in 2000. *Suzaku* represented the fifth X-ray satellite developed by JAXA, and it went aloft atop an M-V-6 rocket from Uchinoura Space Center, Japan. *Suzaku* orbited the Earth in a near-circular pattern at about 345 miles (555 km) altitude.

BELOW
Early returns from *Suzaku*'s X-ray telescopes, which act as X-ray collectors, in contrast to *Chandra*'s telescopes, which are more visually oriented.

BOTTOM
An artist's impression of the *Suzaku* spacecraft, 23 ft (7 m) long, launched into orbit in July 2005.

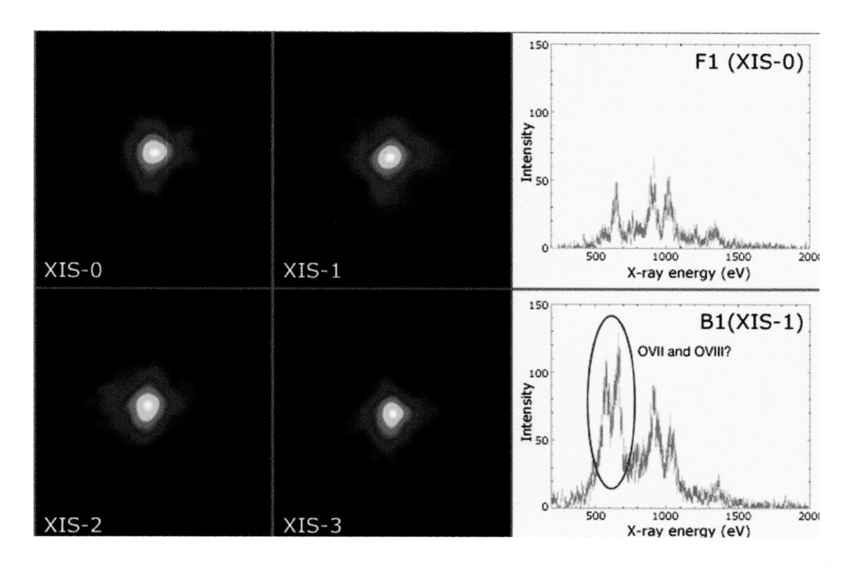

Smaller and lighter than either *Chandra* (see pp. 158–61) or *XMM–Newton* (see pp. 162–63), *Suzaku* still had impressive dimensions and equipment. Some 23 ft (7 m) long with its optical bench extended and 6 ft (1.8 m) in diameter, it weighed about 3500 lb (1590 kg). Its instruments included five X-ray telescopes, four X-ray imaging spectrometers, the Hard X-Ray Detector, and an X-ray spectrometer. The X-ray spectrometer represented a collaboration between JAXA, Tokyo Metropolitan University, NASA Goddard Space Flight Center, and the University of Wisconsin. Its uniqueness stemmed from its means of collecting data—detecting the rise in temperature of a tiny piece of silicon as it absorbs an X-ray photon—and from the accompanying technology (a complicated cryogenic system requiring liquid helium and solid neon in order to reduce temperatures to –460°F/–273°C). This method had the advantage of allowing the measurement of many X-rays, while at the same time determining slight differences in the energy of the incoming photons. The Hard X-Ray Detector, a non-imaging instrument, recorded higher-energy X-rays without the need for a telescope. The four X-ray imaging spectrometers, each with an X-ray-sensitive pixel CCD (similar to the mechanism in a digital camera, but far more sensitive to high-energy light), also involved partnerships between U.S. and Japanese scientific institutions. Finally, the five X-ray telescopes collected light in the form of X-ray photons, then transmitted this data to other instruments, one telescope focusing data on the X-ray spectrometer, and four focusing it on each of the four X-ray imaging spectrometers.

Unfortunately, this project suffered a blow. On August 8, 2005, the helium gas cooling the X-ray spectrometer overwhelmed the technology designed to safeguard it, causing it to vent into space. Without the helium, the intensely cold conditions could not be maintained, and the X-ray spectrometer failed. Mission planners then decided how to maximize the observational science from the remaining instruments for the expected five-year duration of *Suzaku*.

Despite the loss, *Suzaku* continued to provided phenomenal observations with great precision, especially of the mission's main target, the most massive of the universe's black holes. These generally lie at the center of galaxies, compressing the mass of millions, or even billions, of suns into a sphere about the size of our Solar System. Watching the motions of these black holes with unprecedented clarity, scientists could see the influence of intense gravitational forces on space and time, which they observed inferentially by gauging the speed at which black holes spin, the angle at which matter enters the void, and even the evidence of a wall of X-ray light being pulled back and flattened by gravity. More sensitive than *XMM–Newton* in its observations of the same black holes, *Suzaku* allowed scientists not just to guess at the angle at which matter pours into a black hole, but also to actually calculate its angle of descent at about 45 degrees. *Suzaku* also pictured the incredible drama of X-rays escaping from the grip of a black hole, only to be bent back and forced into the sea of debris plummeting into its void.

International Gamma-Ray Astrophysics Laboratory

An artist's drawing of ESA's *INTEGRAL* spacecraft in orbit, which is highly elliptical so that the instruments might avoid the influence of the Earth's radiation.

A representation of how the imaging telescope on board *INTEGRAL* reconstructs pictures of gamma-ray bursts.

Compared to the technological and scientific competition among *Chandra*, *XMM–Newton*, and *Suzaku* (see pp. 158–65), ESA's *International Gamma-Ray Astrophysics Laboratory* (*INTEGRAL*) is conspicuous as a model of cooperative space enterprise. Paradoxically, *INTEGRAL* attempted to observe many of the same phenomena as *XMM–Newton* and *Chandra*: violent and exotic black holes, supernova stars, and neutron stars.

All of the ESA states, as well as the U.S., Russia, the Czech Republic, and Poland, collaborated on *INTEGRAL*. The Italian Space Agency (ASI) acted as the prime contractor, responsible for the design, integration, and testing of the spacecraft. Scientific teams from Italy, Spain, Denmark, Germany, and France contributed the four *INTEGRAL* instruments and oversaw the flow of data. Tracking responsibilities were shared between ESA ground stations and NASA's Deep Space Network operated by JPL. Germany handled satellite control. The Netherlands supervised science operations and Switzerland managed the science data. Twenty-five firms from Europe and one from the U.S. constituted the industrial complement of the project.

Selected for development by ESA in 1993, *INTEGRAL* went into space on October 17, 2002, having lifted off aboard a Proton rocket—Russia's biggest launch vehicle—from Baikonur Cosmodrome in Kazakhstan. The three Proton booster stages placed *INTEGRAL* into Earth orbit, after which its upper stage maneuvered the spacecraft into its transfer orbit, and *INTEGRAL*'s own propulsion system lofted it into its final, highly elliptical pathway (roughly 5592 miles/9000 km from the Earth at its closest point to 95,064 miles/152,991 km at its most distant). This lopsided route enabled long periods of observation uninhibited by the Earth's trapped radiation.

Although smaller than *XMM–Newton*, *INTEGRAL* borrowed features from it and still had significant dimensions and weight (about 16 ft/4.9 m high, 12 ft/3.7 m in diameter, and 52 ft/15.8 m across with the solar panels open, and more than 8820 lb/4000 kg at launch).

The spacecraft consisted of two parts: a lower service module, and a payload module mounted on top of it. The service module represented a re-working of the one on board *XMM–Newton*, a self-contained unit holding the spacecraft's systems. The payload module—the heaviest of its kind to be launched by ESA because of radiation shielding—holds four instruments. Two of them detect gamma rays: an imager providing the sharpest gamma-ray images to date, and a spectrometer, measuring gamma-ray energies. The other instruments (an optical camera and an X-ray monitor) identify the gamma-ray sources.

Gamma-ray bursts, a phenomenon first noticed by military satellites as they scanned for covert nuclear bomb tests, represent one of the mysteries in *INTEGRAL*'s portfolio. They seem to occur at the rate of about one per day. Astronomers think the short bursts result from the collision of neutron stars in distant galaxies, and the longer ones from the explosion of super-massive stars. In addition, *INTEGRAL*'s scientists concern themselves with, among other phenomena, such compact objects as white dwarfs, neutron stars, and potential black holes; the identification of extragalactic high-energy sources; and global surveys involving weekly scans of the galactic plane. Among its more revealing and curious discoveries, *INTEGRAL* observed a neutron star first detected in an active period by NASA's *Rossi X-Ray Timing Explorer* satellite in 1999. In August 2005, *INTEGRAL* "saw" the same star in the central portion of our galaxy, also in an active phase. A month later, with *Rossi* and *INTEGRAL* both trained on it, twenty X-ray bursts were recorded. Realizing that the oscillations accompanying such bursts might relate to the rotations of neutron stars, the scientific team found to their astonishment that this star did indeed revolve—at a rate of about 1122 times per second, which is about as fast as it could spin without flying apart and twice as fast as any such star yet detected.

During December 2006, ESA's Science Program Committee extended *INTEGRAL*'s operations to December 2010.

Key facts

GO-AHEAD DECISION:	**1993**
MANAGEMENT:	**ESA, with contributions by many European countries and the U.S.**
PRIME CONTRACTOR:	**Italian Space Agency**
LAUNCHED:	**October 2002**
PURPOSE:	**Observe gamma-ray bursts in the universe**
ORBIT:	**Highly elliptical**
LENGTH:	**16 ft / 4.9 m**
WEIGHT:	**8820 lb / 4000 kg**
PRIMARY INSTRUMENTS:	**Two gamma-ray detectors, an imager, a spectrometer**
ESTIMATED CESSATION OF OPERATIONS:	**2010**

A striking image taken by *INTEGRAL* in October 2006 of a doughnut-shaped dust cloud encircling a super-massive black hole. ESA has extended *INTEGRAL*'s operations until 2010.

Swift Gamma-Ray Burst Mission

Trained on M101, the *Swift Gamma-Ray Observatory* has learned that many hot, young stars are being formed there, especially along the galaxy's spiral arms.

Key facts

LAUNCHED:	November 2004
MANAGEMENT:	NASA Goddard Space Flight Center
LENGTH:	19 ft / 5.8 m
WEIGHT:	3308 lb / 1500 kg
PURPOSE:	Observe gamma-ray bursts in the universe
PRIMARY INSTRUMENTS:	Three telescopes (gamma-ray, X-ray, and optical spectra)

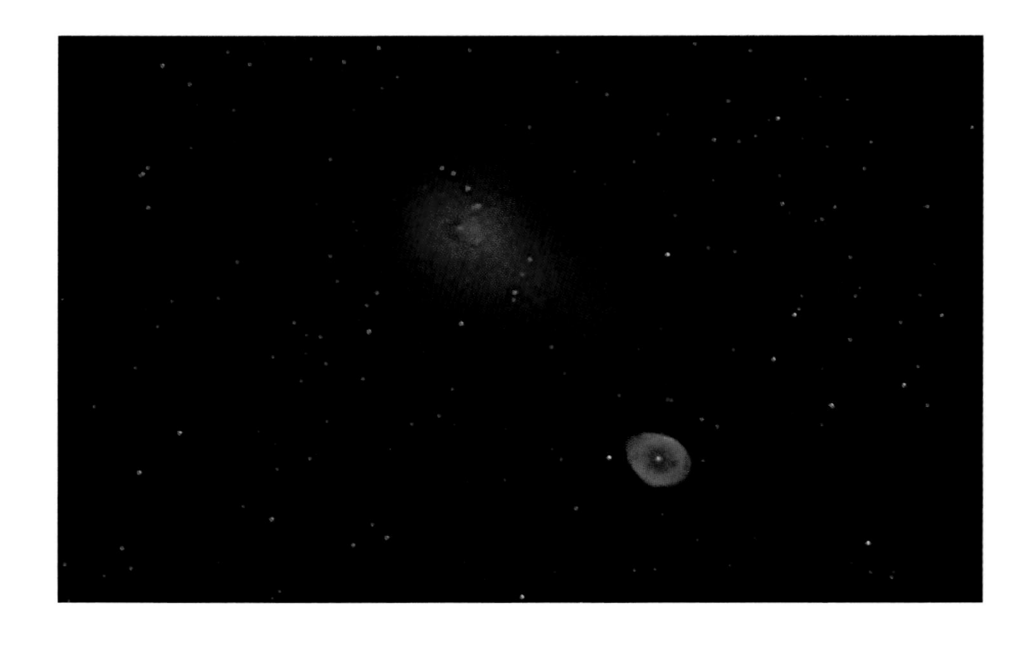

While on the lookout for the Ring Nebula, 12,000 trillion miles (19,312 trillion km) from the Earth, *Swift* inadvertently captured this image of Comet 73P/Schwassmann-Wachmann, some 7.3 million miles (11.7 million km) from our home planet.

In a scenario not unlike the *Chandra/XMM-Newton* duality in X-ray observation, about two years after ESA's *INTEGRAL* went into orbit to chart gamma-ray disturbances in the universe, NASA followed with its own candidate, known as the *Swift Gamma-Ray Burst Mission* ("Swift" refers to the small, agile bird of the same name). Although both projects represented the efforts of many countries and institutions—not just ESA and NASA themselves—in essence *Swift* and *INTEGRAL*, like *Chandra* and *XMM-Newton*, pitted the technology and science of one space agency against that of another.

Swift went into orbit on November 20, 2004, on top of a Delta II launch vehicle that took off from Complex 17A at Cape Canaveral, Florida. Like *INTEGRAL*, *Swift* did have international partners, mainly Italy (the Brera Earth Observatory and the Italian Space Agency Science Data Center), and the U.K. (the Particle Physics and Astronomy Research Council, the Mullard Space Science Laboratory, and the University of Leicester). NASA's Goddard Space Flight Center managed the project. Also like *INTEGRAL*, *Swift* scans the cosmos for violent events in the universe that last from milliseconds to a few minutes. Scientists detect such occurrences from once to several times daily, and they often entail the birth of black holes. Gamma-ray bursts constitute the most powerful explosions in the universe since the Big Bang itself.

In reality, gamma-ray events have attracted the attention of astronomers since the early space age. They first came to light in fall 1963, when the U.S. Air Force—attempting to verify compliance with the U.S.-U.S.S.R. Nuclear Test Ban Treaty—launched the Vela satellites. Later Velas, orbiting in July 1967, recorded the first gamma-ray bursts. Four years later, NASA sent into orbit the *IMP 6* satellite, which also picked up gamma-ray events inadvertently, as it searched for solar flares. NASA then launched the famed *Compton Gamma Ray Observatory* in 1991, which detected 2704 gamma-ray occurrences in nine years of observation. (It re-entered the Earth's atmosphere on June 4, 2000.) The Italian and Dutch space agencies put Europe in the hunt by fabricating and sending into orbit the *BeppoSAX* satellite in 1996. NASA followed *Compton* with MIT's *High Energy Transient Explorer* (*HETE*) in 2000. *INTEGRAL* followed in 2002, and then *Swift*. Finally, NASA has scheduled a 2008 launch of the *Gamma Ray Large Area Space Telescope* (*GLAST*) from Cape Canaveral Air Force Station.

Roughly 19 ft (5.8 m) tall and 18 ft (5.5 m) wide (with solar arrays open), *Swift* weighed about 3308 lb (1500 kg) at launch. *Swift*'s primary mission period began with its liftoff in November 2004 and ended in November 2006; however, it continues to orbit the Earth and is expected to remain operational until 2011.

The *Swift* instrument suite consists of three telescopes, covering gamma-ray, ultraviolet, X-ray, and optical spectra. When the large field-of-view telescope detects an occurrence, the satellite automatically points itself within twenty to seventy-five seconds, so that the two narrow field-of-view telescopes face the eruptions. The large field-of-view telescope—also called the Burst Alert Telescope, or BAT—performs two functions. It serves as the first line of response to a gamma-ray burst, calculating its position. But it also scans the skies, every five minutes accumulating X-ray maps that, in time, will yield a complete survey twenty times more sensitive than the last one, which was conducted during the 1970s. The X-Ray Telescope (XRT) takes images related to the aftereffects of gamma-ray episodes, as well as gathering light energy to determine the chemical elements emitted in explosions. Finally, the Ultraviolet/Optical Telescope (UVOT) captures these events in the visible and ultraviolet ranges, giving physicists another view of the phenomena they wish to understand.

Thus far, *Swift* has recorded many remarkable observations. In July 2005, scientists saw what may have been a flash signifying a neutron star being consumed by a black hole, including the event's afterglow. The following December, they saw a stellar flare on a nearby star of such power that it generated 100 million times more energy than a flare from our own Sun. Then, in May 2007, in one obscure galaxy, *Swift* caught the light of two supernova stars, the first time two of these giants have appeared in one galaxy, exploding at almost the same time.

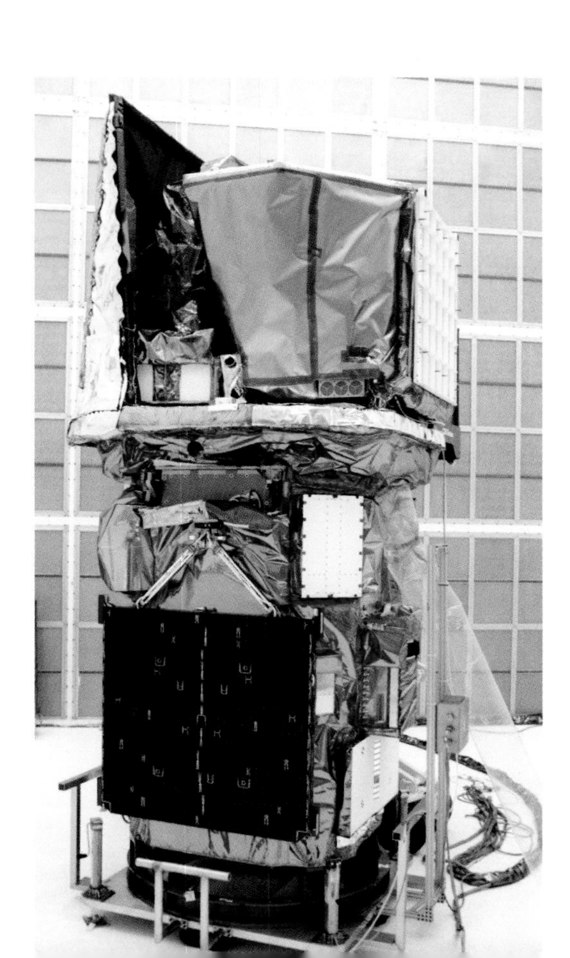

The *Swift* spacecraft sits in a hangar at the Cape Canaveral Air Force Station, just before the removal of its protective cover.

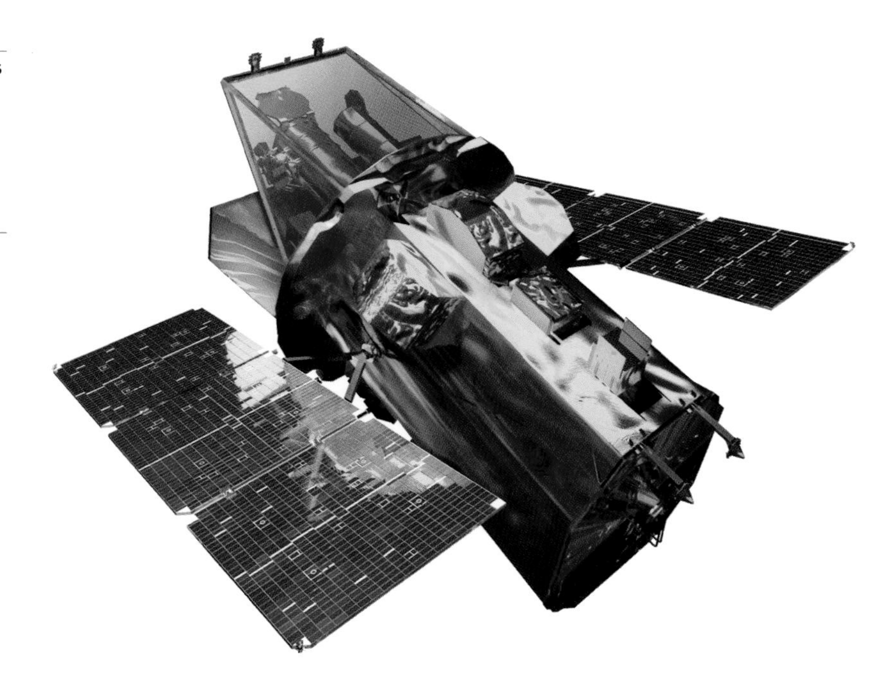

This computer-generated drawing provides a portrait of the diminutive *Swift* spacecraft.

Galaxy Evolution Explorer

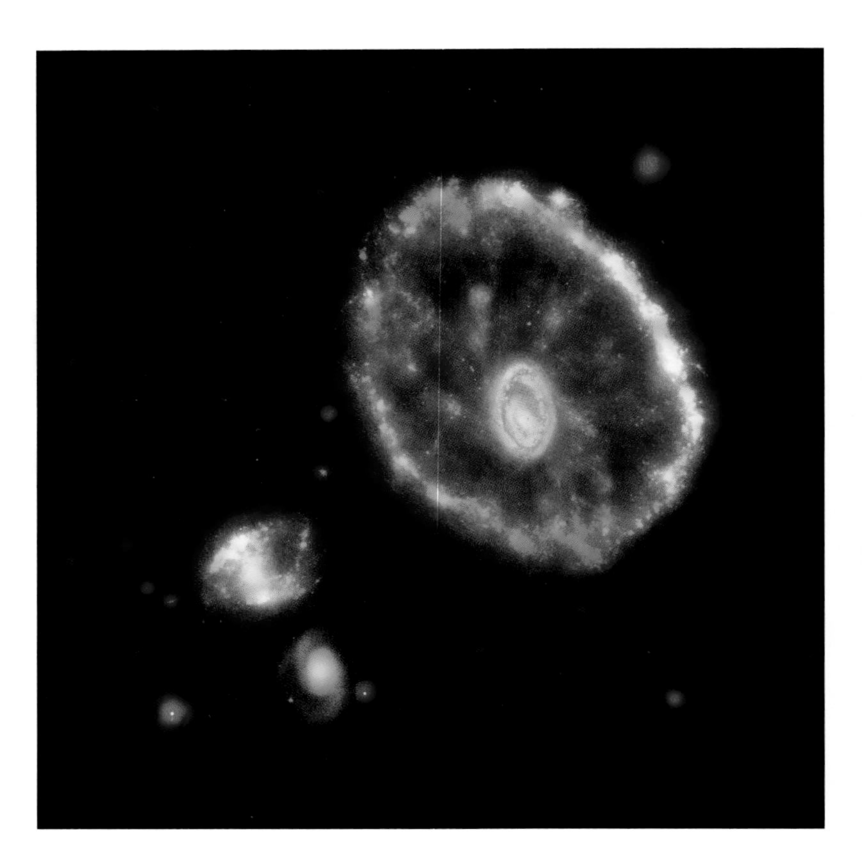

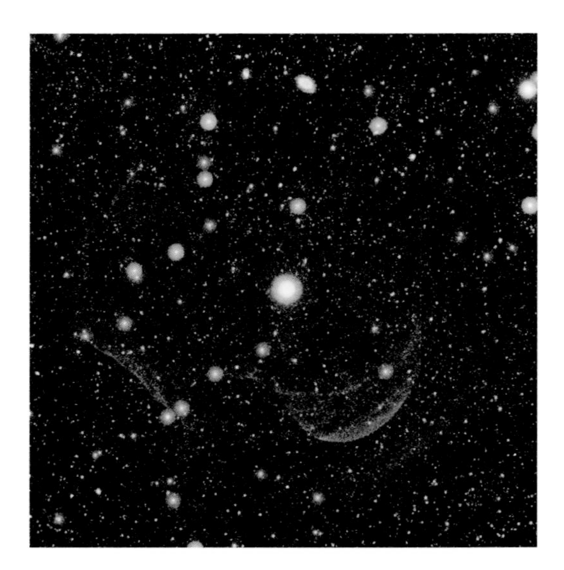

ABOVE, LEFT
This Cartwheel Galaxy picture was produced by four space telescopes: *Chandra*, *Hubble*, *Spitzer*, and *GALEX*, with *GALEX* contributing in the blue range.

ABOVE, RIGHT
The *GALEX* ultraviolet telescope captured this image of the "one-way" galaxy NGC 1512, which evidently grows in size as a result of interaction with other bodies.

LEFT
Capable of watching its subjects over extended periods, *GALEX* first observed the Z Camelopardalis binary star in 2003, but this image dates from early 2004, and was taken in the near and far ultraviolet ranges.

Among such giants in stature as the *Hubble* (see pp. 144–49) and *Chandra* (see pp. 158–61) telescopes, the *Galaxy Evolution Explorer* (*GALEX*) telescope and satellite looked like a dwarf—but a dwarf with imposing credentials.

At launch, *GALEX* weighed a mere 609 lb (276 kg) because of its mostly aluminum structure, and at only 3 ft (0.9 m) wide and 6 ft 4 in. (1.9 m) tall, it could have sat on a sturdy desk. Accordingly, the satellite went aloft not on the back of a big booster, but by design aboard a winged, three-stage, solid-propellant Pegasus XL launch vehicle. Released at 39,000 ft (11.9 km) from an L-1011 aircraft, the Pegasus ignited, driving the *GALEX* out of the atmosphere. These events occurred on April 28, 2003, after the airliner took off from Cape Canaveral Air Force Station, in Florida, and made the drop over the Atlantic Ocean. In ten minutes, the Pegasus inserted *GALEX* into an orbit 428 miles (689 km) over the Earth. Orbital Sciences Corporation, a private firm that conceived and has employed Pegasus for many government and commercial payloads since the 1990s, handled the launch portion of the *GALEX* mission.

A NASA-funded mission managed by JPL, *GALEX* germinated in the Astrophysics Department of the California Institute of Technology (Caltech), who also managed science operations, mission planning,

Key facts

LAUNCHED:	**April 2003 (from an L-1011 aircraft)**
MANAGEMENT:	**NASA Jet Propulsion Laboratory / Caltech**
WEIGHT:	**609 lb / 276 kg**
LENGTH:	**6 ft 4 in / 1.9 m**
PRIMARY INSTRUMENT:	**Telescope with 19¾-in. / 50-cm and 8¾-in. / 22-cm mirrors**
PURPOSE:	**Make sweeping ultraviolet observations of the universe**

and data collection, and supplied the onboard instrument. In addition to launching *GALEX*, Orbital Sciences fabricated the bus (the supporting part of the spacecraft required for the instrument to function, to collect data, and to transmit it) and ran the project's Mission Operations Center. French and South Korean technologies likewise contributed to *GALEX*.

Paradoxically, *GALEX*'s main instrument—a telescope with two mirrors 19¾ in. (50 cm) and 8¾ in. (22 cm) in diameter, respectively—bears a strong resemblance to the design of the giant *HST*. Capable of collecting only one-twentieth of the light of the *HST*, *GALEX* sweeps the cosmos broadly (at 1¼ degrees, or about three times the diameter of the Moon). In contrast, *HST* concentrates on a narrower field-of-view in exquisite detail. But both instruments follow a time-honored pattern, first discovered in the eighteenth century. As light enters the telescope, it travels to the main mirror at the rear, undergoes reflection on to a smaller mirror near the center, and is reflected back to the primary mirror, where it passes through a hole at the mirror's center. From there the light proceeds to a sensor that records the image. In the case of *GALEX*, a lens behind the primary mirror splits the incoming light into two pathways—one deflected to a detector for far-ultraviolet light, the other guided through a lens for the near-ultraviolet spectrum.

Despite its diminutive size, *GALEX* has a formidable mission. From its high vantage point above the Earth, it surveys hundreds of galaxies with each glance—perhaps a million galaxies in all, across 10 billion years of the cosmic past during the spacecraft's lifetime—in order to help astronomers make judgments about the origins of the universe. Originally intended for twenty-eight months of service (which it surpassed in August 2005), *GALEX* continues to observe the cosmos through the ultraviolet spectrum, a range in which the star factories of the universe may be observed—those galaxies dominated by new, hot, short-lived stars. Astronomers believe that the universe originated after the Big Bang about 13.7 billion years ago, and that the creation of galaxies occurred as the subsequent blast of hydrogen and helium expanded and cooled. *GALEX* throws light on the formation of stars, attempting to test the hypothesis that their manufacture may have peaked from 8 to 10 billion years ago. In so doing, it will enable researchers to penetrate 80 percent of the way back to the Big Bang. From this satellite, scientists hope to learn what initiates the process of star formation within galaxies, at what rate they evolve from birth to death, and how many chemical elements form during the life of stars.

In the end, *GALEX* will have created the first map of the ultraviolet universe. In the meantime, the spacecraft has already made some surprising discoveries. For instance, astronomers found confirmation that double, or binary, stars do not seem to experience either massive explosions or occasional, smaller flare-ups, but both; the smaller outbursts seem to happen every few weeks, punctuated by huge explosions every 10,000 years or so. Near its fourth birthday in 2007, *GALEX* recorded the prototypical picture for its creators: a beautiful spiral galaxy, teeming with a swirl of intense, new stars.

Technicians mount the *GALEX* spacecraft inside the cramped payload bay of the small Pegasus rocket, wrapping its solar panels around it.

The *GALEX* satellite, mated to the Pegasus launch vehicle, about to be installed on board an L-1011 mother ship in preparation for its mid-air launch.

An L-1011 aircraft takes off from Cape Canaveral Air Force Station, Florida, with the *GALEX*–Pegasus stack slung under its belly.

Spitzer Space Telescope

The *Spitzer Space Telescope*, one of the four great NASA observatories in space, under construction at Lockheed Martin Space Systems.

The *Spitzer* spacecraft (seen through the imagination of an artist) in a heliocentric orbit, in which it follows the Earth in its movements around the Sun, instead of orbiting the home planet.

Key facts

MANAGEMENT:	**NASA Jet Propulsion Laboratory / Caltech**
LAUNCHED:	**August 2003**
PRIMARY INSTRUMENT:	**Telescope with 33-in. / 84-cm mirror**
LENGTH:	**13 ft / 4 m**
WEIGHT:	**2095 lb / 950 kg**
PURPOSE:	**Make infrared observations of the universe**

The *Spitzer Space Telescope* represents the fourth of NASA's Great Observatories, the other three being *Hubble* (see pp. 144–49), *Chandra* (see pp. 158–61), and the defunct *Compton Gamma Ray Observatory* (see p. 169). As with *GALEX*, JPL managed the project, while Caltech handled the science operations.

Similar to gamma-ray research in space, infrared investigation—*Spitzer*'s specialty—dates from the 1960s. It began with balloon-borne telescopes that made observations above the atmosphere. Then, during the early 1980s, the American, British, and Dutch space agencies collaborated on the world's first infrared telescope, the *Infrared Astronomical Satellite* (*IRAS*). It proved its value over a ten-month mission. Following this foray, the Shuttle Infrared Telescope Facility (SIRTF) flew with *Spacelab 2* in July 1985. Although the Space Shuttle itself contaminated the infrared emissions data during the flight, the mission did succeed in proving the feasibility of a cryogenically cooled telescope in microgravity conditions. Perhaps the most important event in the history of infrared astronomy occurred in 1989, when eminent astronomers and astrophysicists issued a National Research Council report urging the high-priority pursuit of a great infrared observatory in order to delve into some of the most tantalizing problems in astrophysics.

NASA at first responded with visions of a mighty telescope costing some $2.2 billion. Owing to budgetary pressures during the 1990s, however, funding levels fell to the $500 million range. Finally launched atop a Delta II rocket from Cape Canaveral Air Force Station, Florida, on August 25, 2003, the new instrument bore a purely descriptive name, the Space Infrared Telescope Facility. In December of that year, after a public contest, NASA renamed it the *Spitzer Space Telescope* in honor of Professor Lyman Spitzer, Jr. (1914–1997), a renowned astrophysicist at Princeton University, where from 1951 to 1967 he headed the Plasma Physics Laboratory. Spitzer has been credited with being the first—perhaps even preceding Fred Whipple—to call for a space-based telescope, possibly as early as the 1940s. During the 1970s, he championed the *HST* before the U.S. Congress, and deserves at least partial credit for its adoption.

Relatively tall (13 ft/4 m) but lightweight (2095 lb/950 kg, even with cryogen) because of its beryllium structure, *Spitzer* as originally conceived began with a two-and-a-half-year mission length, but in the end its designers expected five years of service. Because infrared involves heat radiation, and the telescope itself could emit infrared that would interfere with sensitive celestial readings, it needed to be kept extremely cold, at −450°F (−268°C). Engineers achieved this feat not by the typical method of infrared protection—total encasement with cryogen inside—but instead by cooling only the instrument chamber at launch, and then subjecting the whole system to the frigidity of space for five weeks prior to operation, in addition to the vapor from its cryogen fluid. As a consequence, planners made drastic reductions in weight and cost, and added greatly to the longevity of the mission (by preserving the cryogen that did fly along).

LEFT
***Spitzer* is mounted inside half of its clam-shell launch fairing, which in turn is situated atop the Delta launch vehicle.**

RIGHT
A technician at Lockheed Martin Space Systems completes integration work on *Spitzer*.

The combined observations of the *Spitzer* (in red), *Chandra* (green and blue), and *Hubble* (yellow) telescopes reveal the full dimensions of the cataclysm that befell the supernova Cassiopeia A.

ABOVE
Spitzer's infrared vision, seen in red, was combined with *Hubble*'s visible-light observations, in blue/green, to produce this phantasm of a merging double galaxy.

ABOVE, RIGHT
Messier 81 looks like the archetypal spiral galaxy in this image from the *Spitzer* library.

RIGHT
The *Spitzer Space Telescope*'s Infrared Array Camera captured this image of a jewel-like spiral galaxy known as NGC 1566.

The spacecraft houses a 33-in. (84-cm) telescope coupled with three science instruments that feature large-format infrared detector arrays. Its orbit—not circling the Earth, but trailing behind it on its trip around the Sun—enables the telescope to operate at a distance far enough from our planet to avoid much of its infrared interference. Moving in this unique pattern, *Spitzer* examines the nearby and the far away. Nearby, it will search for telltale dust disks around neighboring stars, signs of possible planetary formation, and peer behind veils of cosmic dust to see the birth pangs of other stars. Far away, *Spitzer* will look for the black holes or galaxy collisions that spawn ultra-luminous infrared galaxies, and probe the creation of early and distant galaxies. One such discovery involved polycyclic aromatic hydrocarbons, thought to be among the fundamental ingredients of life. Researchers reviewing *Spitzer* data found that these organic molecules could even survive the fury of exploding supernovas, suggesting that in addition to being essential to the formation of life, they are also unimaginably durable.

ABOVE
The Eagle Nebula—an incubator of stars in the Serpens Constellation—is seen here in infrared, courtesy of *Spitzer*.

BELOW
The Helix Nebula, filled with gaseous shells and disks expelled by a dying star, caught the eye of the *Spitzer Space Telescope* in 2006.

RIGHT
The Trifid Nebula in the constellation Sagittarius acts as a gigantic star-making cloud of dust and gas, captured here in all of its intensity by *Spitzer*.

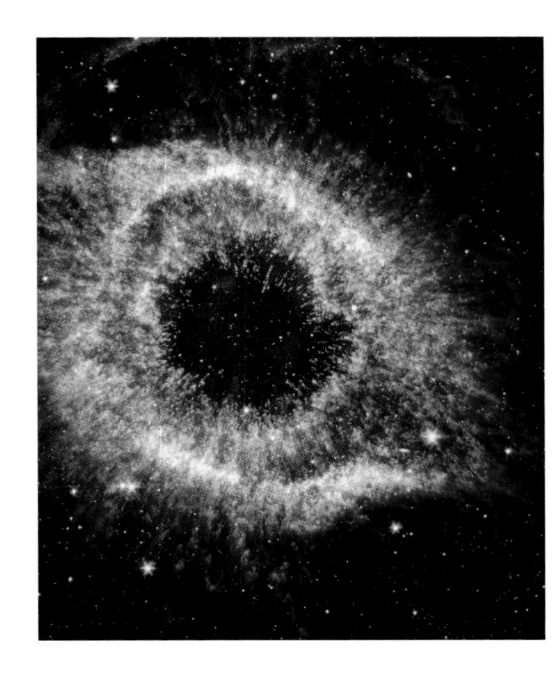

ABOVE
***Spitzer*'s infrared sight contributed the red portions of this image of the M106 spiral galaxy.**

***Spitzer* presented this view of the Pleiades star cluster in 2007, showing the so-called Seven Sisters, shrouded in a veil of dust.**

Epilogue
The Past Revisited

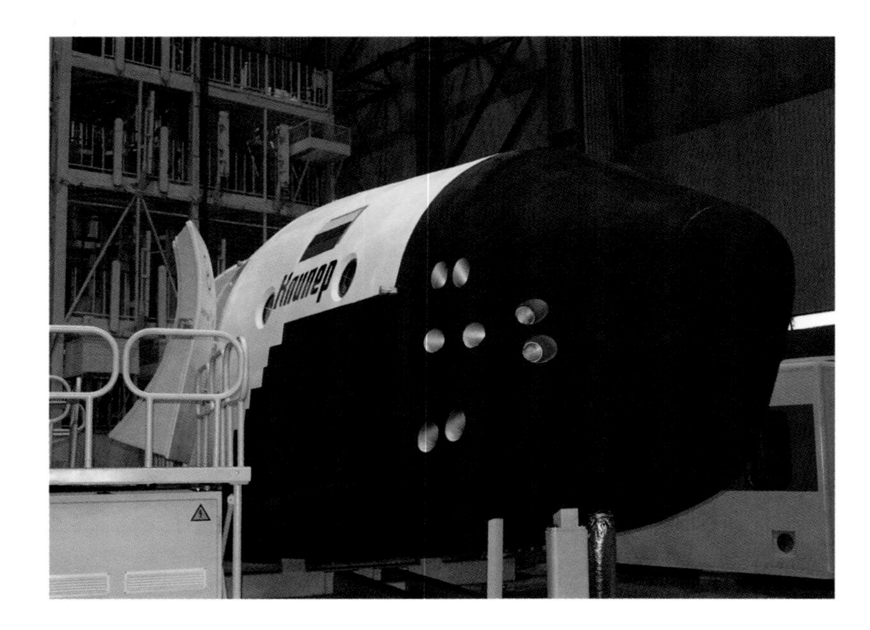

TOP
A mock-up of Kliper, the proposed next-generation Russian spacecraft, to which the Russian Federal Space Agency has expressed a commitment.

ABOVE
The planned Russian Kliper spacecraft (foreground) with Buran, an unfinished project of the past.

Counterintuitive though it may seem, many of the banner activities under development by the world's space powers—especially those involving human spaceflight—seem vaguely familiar, perhaps more traditional than futuristic. Looking ahead to projections by these nations, there appear to be few radically new projects on the horizon. This peculiar conservatism—which, for instance, contemplates a return to the Moon, but no certain settlement there—stems from the unavoidable fact that civil space programs rely on the national treasuries of the countries that pursue them. Aside from the success or failure of the space activities themselves, which certainly have a bearing on the extent of their popular support, all are subject in one form or another to political vagaries. Of course, greater public acceptance and satisfaction should translate into more political leverage for space agencies, and consequently into fatter purses. But because space and politics co-mingle, public approval and generous funding do not necessarily balance one another, and ultimately space programs must compete for money. Their success can depend on a kaleidoscope of broad economic factors, such as the prosperity of the country (or lack of it), budget deficits, and tax policy, to name but a few. Political faction may also play a role as parties make space initiatives more or less of a priority. But most importantly, space programs must fight for a place among all of the essentials that nations pursue: defense, the environment, pensions, medical care, social services, and so forth. As a consequence, civil space programs require not only the support of senior national figures, but also the leadership of space agency officials as capable of wise political judgment as they are of taking sound technical decisions.

In the context of limited and often unpredictable resources, three countries—Russia, China, and the U.S.—seemed inclined toward pursuing the complete array of space activities in the coming decades, including human spaceflight, Earth observation, Solar System exploration, and fundamental space science. The Russian Federal Space Agency (RKA, or "Roskosmos") continues to be a formidable space power, capable in all four areas of activity. Although its plans often appear changeable, its director, Anatoly Perminov, announced decisively in December 2006 that Roskosmos would not join in international efforts to return human beings to the Moon, but would instead develop robotic exploration technologies on its own. The budget for Roskosmos has also been discussed openly: for the ten years between 2006 and 2015, the Russian Duma appropriated about $11 billion. The trend in funding seems to be going up: the 2006 budget of roughly $900 million indicates a rise of 25 percent from 2005, and increases of from 5 to 10 percent per year have been incorporated into the ten-year budget. Additional revenue has been generated from commercial space launches, industrial investments, and space tourism. Still, to give some standard of comparison, in the fiscal year 2007 alone, the U.S. Congress appropriated nearly $17 billion for NASA.

Despite the limitations implicit in their budget, Russian planners remain ambitious. They have said (although doubters abound) that they will develop a successor to Soyuz in the winged, reusable spacecraft called Kliper, hopefully with ESA and JAXA participation, but if necessary without it. Steps have also been taken to modernize and expand both the Soyuz spacecraft (by increasing the number of passengers it can carry from three to four) and the Soyuz rocket (by building Soyuz-2 and Soyuz-2-3, each with greater lifting capacity than the original's 15,000 lb/ 6800 kg). Roskosmos still flies the powerful Proton rocket, with its 40,000-lb (18,140-kg) payload, and its engineers have a new booster

An artist's impression of the Crew Exploration Vehicle, or Orion, the proposed replacement for the Space Shuttle, on an important mission to re-supply the International Space Station with human beings and cargo.

In an artist's rendering, Orion and its crew leave the Earth on board the Ares I launch vehicle. Materials, supplies, and other wherewithal for missions will follow on the giant Ares V booster.

called Angara under development. Lest the importance of the Russian space program be doubted, nearly half of the world's commercial satellites in 2005 went into orbit aboard Roskosmos launch vehicles, and Russia retains a vigorous presence in all aspects of human spaceflight.

The China National Space Agency (CNSA) appears to be well funded and eager to press forward. Cloaked in secrecy, the first major Chinese space initiative may have been started as early as 1968. Between 2001 and 2006, China appears to have launched twenty-two types of satellites, categorized in six series: remote sensing, telecommunications and broadcasting, meteorology, scientific research, Earth resources, and navigation/positioning. It has pursued incremental development of the successful "Long March" rockets, derived from long-range missiles, and has experimented with a new generation of launch vehicles. More importantly, beginning in November 1999, CNSA launched four Shenzhou spacecraft with animals and test dummy passengers, culminating on October 15 and 16, 2003, with a flight by Yang Liwei aboard the *Shenzhou 5* spacecraft. He remained in orbit for twenty-one hours. With this mission, China joined the U.S. and Russia as the only powers to have launched humans into space. China followed this feat with a two-man, five-day mission from October 12 to 17, 2005, aboard the *Shenzhou 6* vehicle. More daring projects have been promised, including extravehicular, rendezvous, and docking maneuvers with astronauts; research aboard short-duration human and long-term autonomous laboratories; the launch of lunar probes and satellites; and the development of technologies for lunar exploration, presumably both mechanical and human. But because the Chinese program plans and pursues its goals in a largely secretive fashion, its future direction remains unclear to outsiders. However, this much is known: on October 24, 2007, the Chinese space agency launched a successful lunar probe, which orbits the Moon and takes three-dimensional images of the surface.

For better or worse—depending on the outcome—the intentions of the U.S. space program became clear to all in 2004. President George W. Bush appeared at NASA Headquarters on January 14 of that year and announced a timetable for a set of major American space initiatives. Even with the memory of the *Columbia* accident the year before still fresh, the president—perhaps emboldened by the successful landing of the Mars Exploration Rover *Spirit* a few days earlier, on January 4—made a daring statement: entitled the "Vision for Space Exploration," the president's plan pledged to complete the International Space Station (ISS) in 2010 and, with its work finished on ISS construction, to end the service life of the Space Shuttle in the same year. President Bush then called for the design and fabrication of a new spacecraft to supplant the Shuttle, capable not only of servicing the ISS, but also of flying outside of Earth orbit to other worlds. He proposed that the spacecraft, known as the Crew Exploration Vehicle and subsequently called Orion, would be ready for its first mission in 2014, and that it would attempt a mission to the Moon no later than 2020. President Bush spoke expansively about the Moon as a future launching platform for Mars and points beyond.

Over time, the Vision for Space Exploration evolved into a well-articulated program. NASA planners added the concept of a partnership among human and robotic sojourners, a worthwhile attempt to move away from the traditional dichotomy of human versus mechanical exploration. In this formulation, the machines go first to do preliminary research, followed by in-depth human voyages of discovery later on.

But as practical hardware concepts developed at NASA and in industry, it became clear that ever-present budgetary pressures foreordained a future in space that resembled the past. In terms of design, Orion strongly resembled the Apollo Command and Service modules, only bigger. Orion as planned would be able to carry six crewmembers to the ISS, and four to the Moon. On lunar voyages, a lander would separate from the Command Module, touch down, and

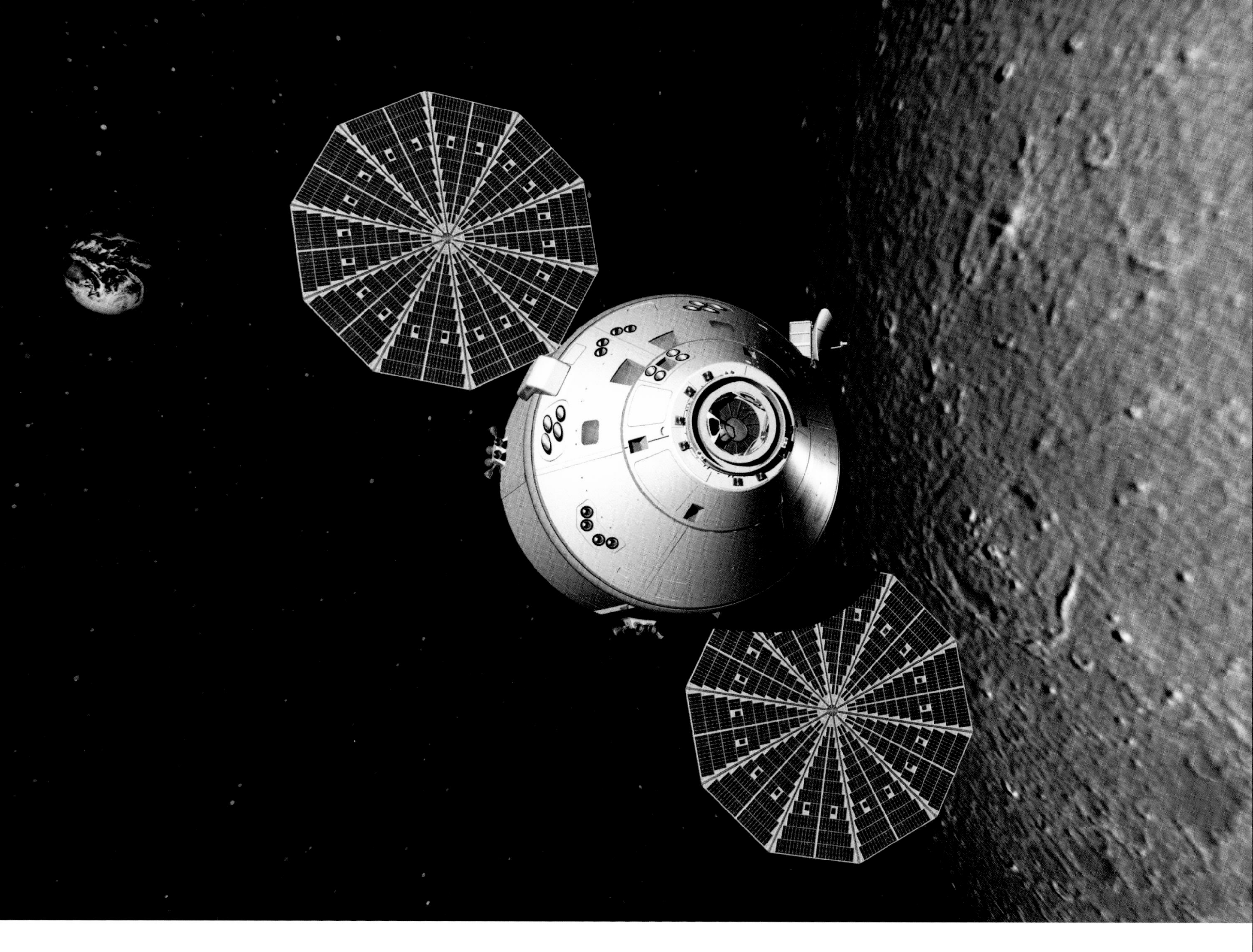

America's planned replacement for the Space Shuttle Orbiter is Orion, shown in this illustration orbiting the Moon.

Orion and the lunar lander orbit the Moon in this artist's conception.

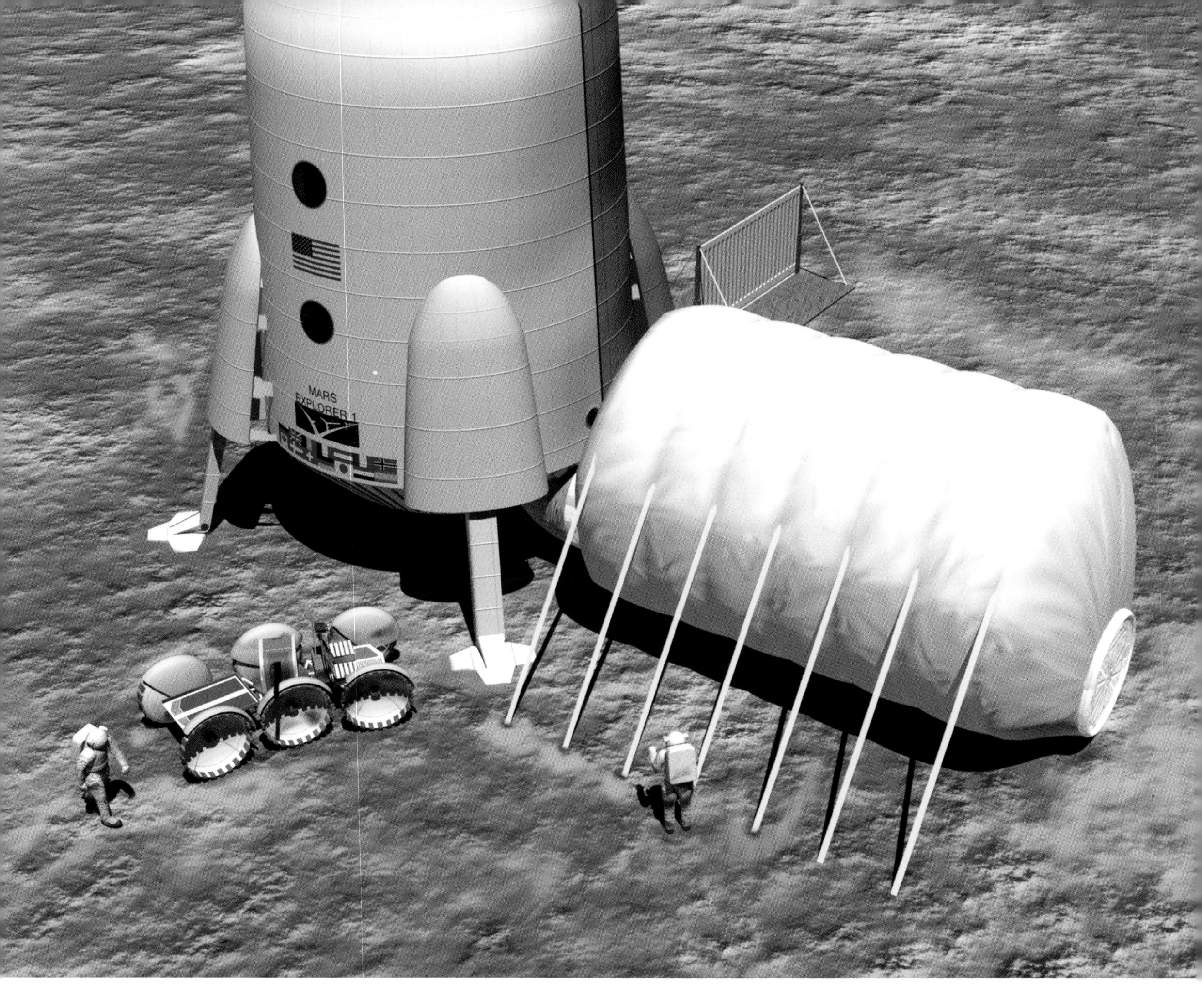

A proposed habitat and airlock for living on Mars, delivered to NASA Langley for ground evaluation.

An artist's impression of a Mars outpost, including a two-story habitat, an inflatable laboratory, and an unpressurized rover vehicle.

remain on the surface for up to seven days before its upper stage took the astronauts back to the orbiting Orion. In addition to the Command Module, Orion also consisted of a Service Module, holding fluids, propulsion systems, and electrical power; a spacecraft adapter to mate it to the launch vehicle; and, above the crew portion, an abort system to enable escape in case of launch emergencies. The planned return to Earth from space likewise seemed familiar—jettisoning the Service Module and falling through the atmosphere on parachutes—but differed in that the Command Module, equipped with airbag cushions and retrorockets, would land on the ground somewhere in the western U.S. After a seven-month competition for Orion between Lockheed Martin and Northrop-Grumman/Boeing, Lockheed won the prime contract on August 31, 2006.

If Orion seemed like "Apollo on steroids," as NASA Administrator Michael Griffin jokingly called it, the boost vehicle designed to carry it into orbit borrowed liberally from the rocketry systems used on the Space Shuttle, as well as from the Saturn launch vehicles. Known as Ares, it will actually be made up of two separate boosters (Ares I and Ares V, in tribute to Saturn I and V). Orion would rise to Low Earth Orbit aboard Ares I, which will consist of a single, five-segment solid rocket booster, derived from those on the Shuttle, and a second-stage liquid oxygen/hydrogen engine based on the Saturn V's J-2 second-stage engines. The massive Ares V—heavier even than the Saturn V and capable of lifting more than 286,000 lb (129,700 kg) into orbit—also drew inspiration from the Shuttle. It will be powered in its first stage by five RS-68 liquid oxygen/hydrogen engines installed under an external tank bigger than that used for the Shuttle, as well as two of the solid rocket boosters used on Ares I. The second stage will be the same as that on Ares I. The Ares V will deliver supplies, equipment, and components into orbit, where Orion will then be equipped as necessary for its voyages to the Moon and, if it happens, to Mars. In July 2007, NASA awarded a $1.2 billion contract to Pratt and Whitney Rocketdyne to fabricate the Ares upper stages.

Known in its totality by the name Constellation, this grand design for American spaceflight might cost the U.S. taxpayer about $104 billion in total—no small price, despite the project's retrospective features.

An illustration of the lunar lander on the Moon's surface, with the Earth above the horizon.

PAGES 184–85
A drawing of the Ares V launch vehicle's Earth Departure Stage (left), propelling the Lunar Surface Access Module (center) and the mated Orion (right) toward the Moon.

Notes

Chapter 1

1 National Aeronautics and Space Administration, *Columbia Accident Investigation Board Report*, vol. 1, Washington, D.C. (Government Printing Office) 2003, p. 25.
2 "Flight History of Canadarm" (STS-2 entry), Canadian Space Agency, space.gc.ca/asc/eng/exploration/canadarm/flight.asp?printer=1, accessed February 7, 2008.
3 "Canadarm Introduction," Canadian Space Agency, space.gc.ca/asc/eng/exploration/canadarm/introduction.asp, updated November 21, 2001, accessed February 7, 2008.
4 "Collocation of ESA ATV Management Team and Prime Contractor EADS-LV: A Very Positive Step," ESA News, esa.int/SPECIALS/ATV/ESADL6OED2D_2.html, dated June 29, 2002, accessed February 7, 2008.

Chapter 2

1 Owing to space limitations, neither PARASOL nor the Orbiting Carbon Observatory (a JPL satellite) can be covered in this book. Lidar is an acronym of "Light Detection and Ranging."
2 "EO-1/SAC-C Spacecraft Successfully Launched," NASA EO News, earthobservatory.nasa.gov/Newsroom/NasaNews/2000/200011214247.html, dated November 21, 2000, accessed February 7, 2008.

Chapter 4

1 "XMM-Newton Overview," ESA, esa.int/esaSC/120385_index_2_m.html, updated October 14, 2003, accessed February 7, 2008 (emphasis added).

Acronyms

AEB Brazilian Space Agency (Agência Espacial Brasileira)
AIRS Atmospheric Infrared Sounder
ALICE A Lightweight Imaging Spectrometer for Cometary Exploration
AMSR-E Advanced Microwave Scanning Radiometer-EOS
AMSU Advanced Microwave Sounding Unit
APXS Alpha Particle X-Ray Spectrometer
ASI Italian Space Agency (Agenzia Spaziale Italiana)
ASPOC Active Spacecraft Potential Control
ASTER Advanced Spaceborne Thermal Emission and Reflection Radiometer
ASTP Apollo-Soyuz Test Project
ATV Automated Transfer Vehicle
AXAF Advanced X-Ray Astrophysics Facility
BAT Burst Alert Telescope
CAIB Columbia Accident Investigation Board
CALIOP Cloud-Aerosol Lidar with Orthogonal Polarization
CALIPSO Cloud-Aerosol Lidar and Infrared Pathfinder and Satellite Observation
Caltech California Institute of Technology
CCD Charge Coupled Device
CDS Coronal Diagnostics Spectrometer
CELIAS Charge, Element, and Isotope Analysis System
CERES Clouds and the Earth's Radiant Energy System
CIS Cluster Ion Spectrometry
CNES Centre National d'Études Spatiales
CNSA China National Space Agency
CONAE Argentinian National Space Activities Commission (Comisión Nacional de Actividades Espaciales)
COSTEP Comprehensive Suprathermal and Energetic Particle Analyzer
CPR Cloud Profiling Radar
CSA Canadian Space Agency
DORIS Doppler Orbitography and Radiopositioning Integrated by Satellite
DSRI Danish Space Research Institute (Dansk Rumsforskningsinstitut)
DWP Digital Wave Processing
EADS European Aeronautic Defence and Space Company
EDI Electron-Drift Instrument
EFW Electric Field and Wave
EIT Extreme Ultraviolet Imaging Telescope
EOS Earth Observing System
EOSAT Earth Observation Satellite
EPA Environmental Protection Agency
ERNE Energetic and Relativistic Nuclei and Electron experiment
ERTS Earth Resources Technology Satellites
ESA European Space Agency
ETM+ Enhanced Thematic Mapper Plus
EUMETSAT European Organisation for the Exploitation of Meteorological Satellites
EVA Extravehicular Activity
FGB Functional Cargo Block (*Funktsional'no-Gruzovoi Blok*)
FGM Fluxgate Magnetometer
FUSE Far Ultraviolet Spectroscopic Explorer
GALEX Galaxy Evolution Explorer
GLAST Gamma Ray Large Area Space Telescope
GOES Geostationary Operational Environmental Satellites
GOLF Global Oscillations at Low Frequencies
GOLPE GPS Occultation and Passive Reflection Experiment
GPS Global Positioning System
GRACE Gravity Recovery and Climate Experiment
HAIRS High Accuracy Intersatellite Ranging System
HEND High Energy Neutron Detector
HETE High Energy Transient Explorer
HIRDLS High Resolution Dynamics Limb Sounder
HRTC High Resolution Technological Camera
HSB Humidity Sounder for Brazil
HSC High Sensitivity Camera
HST Hubble Space Telescope
IAU International Astronomical Union
IIR Imaging Infrared Radiometer
INTEGRAL International Gamma-Ray Astrophysics Laboratory
IRAS Infrared Astronomical Satellite
ISS International Space Station
IST Italian Star Tracker
IUS Inertial Upper Stage
JAXA Japanese Aerospace Exploration Agency
JEM Japanese Experiment Module
JMR Jason Microwave Radiometer
JPL Jet Propulsion Laboratory (NASA)
LASCO Large Angle Spectrometric Coronagraph experiment
Lidar Light Detection and Ranging
LIS Lightning Image Sensor
LRRA Laser Retroflector Array
MBS Mobile Base System
MDI Michelson Doppler Imager
MELFI Minus Eighty Degree Laboratory Freezer for ISS
MESSENGER Mercury Surface, Space Environment, Geochemistry, and Ranging
Mini-TES Miniature Thermal Emission Spectrometer
MISR Multi-Angle Imaging Spectroradiometer
MIT Massachusetts Institute of Technology
MLS Microwave Limb Sounder
MMP Magnetic Mapping Payload
MMRS Multispectral Medium Resolution Scanner
MODIS Moderate-Resolution Imaging Spectroradiometer
MOPITT Measurements of Pollution in the Troposphere
MPLM Multi-Purpose Logistics Module
MUSES Mu Space Engineering Spacecraft
NASA National Aeronautics and Space Administration
NOAA National Oceanic and Atmospheric Administration
OMI Ozone Monitoring Instrument
OTCMs Orbital Replacement Unit Tool Changeout Mechanisms
PAM Payload Assisted Module
PARASOL Polarization and Anisotropy of Reflectances for Atmospheric Sciences coupled with Observations from a Lidar

PEACE	Plasma Electron and Current Experiment
PEPPSI	Pluto Energetic Particle Spectrometer Science Investigation
PR	Precipitation Radar
RAPID	Research with Adaptive Particle Imaging Detectors
REX	Radio Science Experiment
RKA	Russian Federal Space Agency (Roskosmos)
RTG	Radioisotope Thermoelectric Generator
SAC-C	Satelite de Aplicaciones Cientificas-C (Scientific Applications Satellite)
SEM	Space Environment Monitor
SHM	Scalar Helium Magnetometer
SIM	Spectral Irradiance Monitor
SIRTF	Shuttle Infrared Telescope Facility
SMRS	Space Shuttle Attached Remote Manipulator System
SOHO	Solar and Heliospheric Observatory
SOI	Solar Oscillations Investigation
SOLSTICE	Solar Stellar Irradiance Comparison Experiments
SORCE	Solar Radiation and Climate Experiment
SSN	Space Surveillance Network
STAFF	Spatio-Temporal Analysis of Field Fluctuation
STS	Space Transportation System
SUMER	Solar Ultraviolet Measurements of Emitted Radiation
SWAN	Solar Wind Anistropies
SWAP	Solar Wind Analyzer Around Pluto
TES	Tropospheric Emission Spectrometer
TIM	Total Irradiance Monitor
TIROS	Television Infrared Observation Satellite
TKS	Transport Logistics Spacecraft (*Transportniy Korabl Snabzheniya*)
TMI	TRMM Microwave Imager
TOPEX	The Ocean Topography Experiment
TRMM	Tropical Rainfall Measuring Mission
TSI	Total Solar Irradiance
USGS	United States Geological Survey
UVCS	Ultraviolet Coronagraph Spectrometer
UVOT	Ultraviolet/Optical Telescope
VIM	Voyager Interstellar Mission
VIRGO	Variability of Solar Irradiance and Gravity Oscillations
VIRS	Visible Infrared Scanner
WBD	Wide Band Data
WHISPER	Waves of High Frequency and Sounder for Probing of Electron Density by Relaxation
XMM	X-Ray Multi-Mirror Design
XPS	XUV Photometer System
XRT	X-Ray Telescope

Further Reading

Online resources

This book is largely a product of the Internet. In mining the data for it, a handful of distinguished websites served as indispensable portals to the world of spaceflight. Indeed, it seemed natural to turn to the Internet as the main resource. The project started with a provocatively simple question posed to the author by Julian Honer, Merrell's Editorial Director: What is flying in space right now? It soon became clear that the Internet offered the only way to answer him fully.

The following four websites proved to be invaluable. While they all proclaim their country's successes enthusiastically—sometimes without due regard for the failures, or the successes, of other space programs—their data are vast, thorough, and usually accurate.

- **nasa.gov** The National Aeronautics and Space Administration (see also the superb websites of NASA's field centers, particularly the Jet Propulsion Laboratory, the Goddard Space Flight Center, the Marshall Space Flight Center, the Kennedy Space Center, the Johnson Space Center, the Stennis Space Center, and the Ames Research Center)
- **esa.int** The European Space Agency
- **jaxa.jp** The Japanese Aerospace Exploration Agency
- **space.gc.ca** The Canadian Space Agency

Of course, other websites were consulted, but those listed above constitute the bedrock of the research. Additionally, because many of the projects described in this book involve bilateral, or even multinational, partnerships, these four websites often present the best insights into, and information about, those space programs with which the main space administrations are cooperating.

Selected bibliography

In addition to the Internet, a number of published works have offered necessary sources of reference, technical knowledge, and historical context for the fifty-seven spacecraft described in this book.

Baker, David, ed., *Jane's Space Directory, 2005–2006*, Coulsdon, Surrey, U.K. (Jane's Information Group) 2005

Bilstein, Roger E., *Testing Aircraft, Exploring Space: An Illustrated History of NACA and NASA*, Baltimore and London (The Johns Hopkins University Press) 2003

Bromberg, Joan Lisa, *NASA and the Space Industry*, Baltimore and London (The Johns Hopkins University Press) 1999

Burrows, William E., *The Infinite Journey: Eyewitness Accounts of NASA and the Age of Space*, New York (Discovery Books) 2000

Butrica, Andrew J., *Beyond the Ionosphere: Fifty Years of Satellite Communication*, NASA Special Publication 4217, Washington, D.C. (NASA) 1997

Collins, Martin J. and Sylvia K. Kraemer, eds, *Space: Discovery and Exploration*, New York (Hugh Lauter Levin) 1994

Dethloff, Henry C. and Ronald A. Schorn, *Voyager's Grand Tour: To the Outer Planets and Beyond*, Washington, D.C., and London (Smithsonian Institution Press) 2003

Gorn, Michael H., *NASA: The Complete Illustrated History*, rev. edn, London and New York (Merrell Publishers), 2008

Huffbauer, Karl, *Exploring the Sun: Solar Science Since Galileo*, Baltimore and London (The Johns Hopkins University Press) 1991

Jenkins, Dennis R., *Space Shuttle: A History of the National Space Transportation System*, Stillwater, Minn. (Voyageur Press) 2002

Kelly, Thomas J., *Moon Lander: How We Developed the Apollo Lunar Module*, Washington, D.C., and London (Smithsonian Institution Press) 2001

Kraemer, Robert S., *Beyond the Moon: The Golden Age of Planetary Exploration, 1971–1978*, Washington, D.C., and London (Smithsonian Institution Press) 2000

Krige, John et al., *A History of the European Space Agency, 1958–1987*, 2 vols, ESA Special Publication 1235, Noordwijk, The Netherlands (ESA) 2000

Launius, Roger D., *A History of the U.S. Civil Space Program*, Malabar, Fla. (Krieger Publishing) 1994

Launius, Roger D., and Bertram Ulrich, *NASA and the Exploration of Space: With Works from the NASA Art Collection*, New York (Stewart, Tabori & Chang) 1998

McCurdy, Howard E., *Faster, Better, Cheaper: Low-Cost Innovation in the U.S. Space Program*, Baltimore and London (The Johns Hopkins University Press) 2001

Mudgway, Douglas J., *Uplink-Downlink: A History of the Deep Space Network, 1957–1997*, NASA Special Publication 4227, Washington, D.C. (NASA) 2001

National Aeronautics and Space Administration, *Columbia Accident Investigation Board Report*, vol. 1, Washington, D.C. (Government Printing Office) 2003

Schorn, Ronald M., *Planetary Astronomy: From Ancient Times to the Third Millennium*, College Station, Tex. (Texas A&M University Press) 1998

Siddiqi, Asif, *Challenge to Apollo: The Soviet Union and the Space Race, 1945–1974*, NASA Special Publication 4408, Washington, D.C. (NASA) 2000

Siddiqi, Asif, *Deep Space Chronicle: A Chronology of Deep Space and Planetary Probes, 1958–2000*, NASA Monographs in Aerospace History Number 24, Washington, D.C. (NASA) 2002

Finally, one of the most noteworthy sources for books about the American space program is the NASA website, on which can be found a library of volumes in the fine NASA History Series, most available in full text, all at no charge. To find out more, go to history.nasa.gov/on-line.html.

Acknowledgments/Picture Credits

Writing books is always a cooperative venture, but even more so for illustrated titles.

The concept of this book originated with Merrell's Publisher, Hugh Merrell, and with the Editorial Director, Julian Honer. They conceived the idea of a volume that surveys the pivotal spacecraft of today, and are also responsible for its scope and framework. I am grateful for their creation.

Mark Ralph of Merrell represents the best kind of editor—one who discerns the author's intentions and improves whatever he touches. He was abetted ably by copy-editor Gordon Lee.

Two individuals read this book in its early stages. Noted air and space author Curtis Peebles dissected the manuscript painstakingly, and the final version is much more accurate as a consequence of his labors. In addition, David Baker, one of the world's foremost experts on the global space scene, offered much encouragement.

Merrell Publishers specializes in beautiful books, and thanks to Senior Designer Paul Arnot—and Paul Spencer of Found Design, London—this one is more than equal to the standard. The designers' success is also due to the fine selection of images provided by picture editors Roland and Sarah Smithies of Luped Picture Research.

None of these elements would have jelled without the essential ingredient, Michelle Draycott, Merrell's Production Manager.

But the best help came from home. My wife, Christine Gorn, offered the most generous love and loyalty that anyone could ever hope for.

Michael H. Gorn

The illustrations in this book have been reproduced courtesy of the following copyright holders:

Anatoly Zak/russianspaceweb.com: 178t, 178b. **Ball Aerospace and Technologies Corp.:** 119tl, 119tr. **Boeing:** 85br. **Boeing/Thom Baur:** 78. **Canadian Space Agency:** 56. **CONAE:** 88t, 88b, 89. **Corbis:** 105. **David A. Hardy/ Science Photo Library:** 150. **ESA:** front endpaper, 50l, 50r, 63l, 63r, 92t, 92b, 93l, 93r, 93b, 94 (all), 95, 96, 97, 126b, 127t, 162t, 163tl, 163r, 166t, 166b, 167. **ESA/A. Van Der Geest:** 136bl, 136br, 137l. **ESA/AOES Medialab:** 98–99, 130l, 132–33, 134, 136t, 136cl. **ESA/C. Fuglesang:** 17. **ESA/D. Ducros:** 130r, 51t, 51bl, 51br, 52t, 52b. **ESA/DLR/FU Berlin (G. Neukum):** 131tl, 131r, 131bl. **ESA/EPIC/MPE:** 163bl. **ESA/J.L. Atteleyn:** 135c. **ESA © 2007 MPS for OSIRIS Team MPS/UPD/LAM/IAA/RSSD/ INTA/UPM/DASP/IDA:** 100, 101bl, 101br, 137rt, 137rb. **ESA/NASA:** 101t, 127bl. **ESA/NASA/JPL/University of Arizona:** 129bl. **ESA/S. Corvaja:** 53b 135t. **ESA/Science Photo Library:** 151bl. **ESA/Thomas Reiter:** Back jacket (tl), 29b. **ESA/VIRTIS/INAF-IASF/Obs. de Paris-LESIA:** 135bl, 135br. **ESA/XMM-Newton/A. De Luca (INAF-IASF):** 143c. **ISAS/JAXA/Kousuke Oonuki:** 165b. **JAXA:** 54l, 54r, 55, 139. **John Frassanito and Associates:** 182t, 183. **Julian Baum/Science Photo Library:** 151t. **Lockheed Martin Corp.:** 176–77, 179, 180, 181t, 181b. **Lockheed Martin Space Systems/Russ Underwood:** 172l, 173r. **MD Robotics:** 57. **MEF/ISAS/A. Ikeshita:** 138l, 138t, 138br. **NASA:** Front jacket, back jacket (tr, cr, bl, br), back endpaper, 2–3, 4–5, 6, 7bl, 7br, 8, 9t, 10, 12–13, 15, 16, 18t, 18b, 19, 20, 21l, 21r, 22, 23, 24–25, 26, 27, 28t, 28b, 29t, 30t, 30bl, 30br, 31, 32t, 32b, 33, 34bl, 34br, 35l, 35r, 36–37, 38l, 38r, 39t, 39b, 40tl, 40tc, 40tr, 40b, 41t, 41b, 42t, 42c, 42b, 43tl, 43tr, 43b, 44l, 44tr, 44r, 45t, 45b, 46, 47t, 47c, 47b, 48, 49t, 49bl, 49br, 53t, 60, 61t, 68b, 69, 72, 73l, 74, 75t, 75b, 76l, 76r, 77, 79bl, 80t, 80b, 85t, 91, 103t, 103b, 118t, 118c, 119br, 143t, 144, 145, 146l, 146r, 154, 155, 158, 164t, 164b, 165t, 169l, 171t, 171c, 171b, 173l. **NASA/C.R. O'Dell and S.K. Wong (Rice University):** 149br. **NASA/CXC/ASU/J. Hester *et al.* (X-ray); NASA/ESA/ASU/J. Hesterand A. Loll (Optical); NASA/JPL-Caltech/University of Minnesota/R. Gehrz (Optical):** 160t. **NASA/CXC/Penn State University/ L. Townsley *et al.* (X-ray); Pal. Obs. DSS (Optical):** 160bl. **NASA/CXC/University of Bristol/Worrall *et al.* (X-ray); NRAO/AUI/NSF (Radio):** 161br. **NASA/CXC/University of Colorado/Linsky *et al.* (X-ray); NASA/ESA/STScI/ ASU/J. Hester and P. Scowen (Optical):** 160br. **NASA/CXC/University of Maryland/A.S. Wilson *et al.* (X-ray); Pal. Obs. DSS (Optical); NASA/JPL-Caltech (IR); NRAO/AUI/NSF (VLA):** 161bl. **NASA/E.J. Schreier (STScI):** 149tl. **NASA/EPO/Sonoma State University, Aurore Simonnet:** 169r. **NASA/ESA/A. Nota (STScI/ ESA):** 147. **NASA/ESA/HEIC/The Hubble Heritage Team (AURA/STScI):** 149tr. **NASA/ESA/JPL-Caltech/STScI/ D. Elmegreen (Vassar):** 174cl. **NASA/ESA/R. Massey (Caltech):** 162b. **NASA/ESA/R. Sankrit and W. Blair (JHU):** 149cl. **NASA/ESA/The Hubble Heritage Team (STScI/AURA):** Back jacket (c), 148r, 148cl, 149bl. **NASA/GFSC:** 65tl, 83. **NASA/GSFC/LaRC/JPL, MISR Team:** 61b, 67b. **NASA/GSFC/METI/ERSDAC/JAROS and U.S./Japan ASTER Science Team:** 66b, 67t. **NASA/GSFC/Reto Stöckli/Robert Simmon/MODIS Team/USGS EROS Data Center/USGS Terrestrial Remote Sensing Flagstaff Field Center/Defense Meteorological Satellite Program:** 62. **NASA/GSFC Scientific Visualization Studio:** 58–59, 71t, 90bc, 90br. **NASA/Jeff Caplan:** 182b. **NASA/Jeff Schmaltz, MODIS Rapid Response Team/GSFC:** 71b. **NASA/JHUAPL/ Carnegie Institution of Washington:** 116, 117. **NASA/ JHUAPL/SwRI:** 11, 122, 123 (all). **NASA/Jon Morse (University of Colorado):** 143br. **NASA/JPL:** Back jacket (cl), 64–65, 68t, 73t, 73cr, 79r, 86, 104t, 104bl, 104br, 106tl, 106tr, 106bl, 106br, 107, 109tl, 110, 111, 112–13, 120, 121t, 124t, 124b, 126t, 129cr, 151br. **NASA/JPL/Arizona State University:** 125b. **NASA/JPL/ASU:** 109b. **NASA/JPL/ CIRA, Colorado State University/NOAA:** 79t, 85bl. **NASA/JPL/Cornell:** 114b. **NASA/JPL/Malin Space Science Systems:** 108. **NASA/JPL/NIMA:** 64t. **NASA/ JPL/Space Science Institute:** 128 (all), 129t, 129cl, 129c. **NASA/JPL/Space Science Institute/University of Arizona:** 129br. **NASA/JPL/Texas A&M/Cornell:** 115t, 115cr. **NASA/JPL/UMD, Pat Rawlings:** 118b. **NASA/JPL/University of Arizona:** 121c, 121b. **NASA/JPL/University of Texas CSR:** 81t, 81b. **NASA/JPL-Caltech:** 87, 172r. **NASA/JPL-Caltech/ASU/ Cornell/Don Davis:** 125t. **NASA/JPL-Caltech/Cornell:** 114t, 115cl, 115b. **NASA/JPL-Caltech/DSS/GALEX:** 170tr. **NASA/JPL-Caltech/J. Hora (Harvard-Smithsonian CfA):** 175bl, 175br. **NASA/JPL-Caltech/J. Rho (SSC/Caltech):** 175c. **NASA/JPL-Caltech/J. Stauffer (SSC/Caltech):** 175bc. **NASA/JPL-Caltech/L. Rudnick (University of Minnesota):** 142. **NASA/JPL-Caltech/M. Seibert (OCIW)/T. Pyle (SSC)/R. Hurt (SSC):** 170bl. **NASA/ JPL-Caltech/MSSS:** 109tr, 109cr. **NASA/JPL-Caltech/ N. Flagey (IAS/SSC) and A. Noriega-Crespo (SSC/Caltech):** 174t, 175t. **NASA/JPL-Caltech/ P.N. Appleton (Spitzer Science Center/Caltech):** 170tl. **NASA/JPL-Caltech/R. Kennicutt (University of Arizona) and the SINGS Team:** 174b. **NASA/JPL-Caltech/ S. Willner (Harvard-Smithsonian CfA):** 174cr. **NASA/ JPL-Caltech/The Hubble Heritage Team (STScI/AURA):** 148tl. **NASA/JPL-Caltech/UMD:** 9b, 119bl, 119bc. **NASA/KSC:** 159l, 159r. **NASA/LASP, University of Colorado/Orbital Sciences Corporation:** 82. **NASA/MSFC:** 34t, 34c, 184–85. **NASA/Science Photo Library:** 102l, 102r, 103ct, 103cb, 127br. **NASA/Swift/ UVOT/PSU/Peter Brown:** 168t, 168b. **NASA/The Hubble Heritage Team (AURA/STScI):** 143bl, 148bl. **NASA, Barbara Summey/GSFC:** 66t. **Northrop Grumman/ Sally Aristei:** 70. **SOHO (ESA and NASA):** 152, 153l, 153r. **SSAI/NASA/GSFC:** 90t, 90bl. **T.A. Rector and B.A. Wolpa, NOAO, AURA, NSF:** 140–41, 156–57. **USGS:** 65br.

a: above
b: below
c: center
l: left
r: right
t: top

Index

Figures in *italics* refer to illustrations.
For acronyms, see pp. 186–87.

Once again, for Christine
The light of my life

First published 2008 by Merrell Publishers Limited

Head office:
81 Southwark Street
London SE1 0HX

New York office:
740 Broadway, Suite 1202
New York, NY 10003

merrellpublishers.com

A catalog record for this book is available from the Library of Congress.

British Library Cataloguing-in-Publication data:
Gorn, Michael H.
Superstructures in space : from satellites to space stations – a guide to what's out there
1. Artificial satellites
I. Title
629.4'6

ISBN-13: 978-1-8589-4417-3
ISBN-10: 1-8589-4417-1

Produced by Merrell Publishers Limited
Designed by Paul Spencer at Found Design Limited
Copy-edited by Gordon Lee
Proof-read by Elizabeth Tatham
Indexed by Vicki Robinson

Printed and bound in China

Front jacket: This magnificent photograph was taken by an International Space Station (ISS) Expedition 15 crewmember just after the Space Shuttle *Atlantis* (STS-117) had undocked from the ISS in June 2007. Pictured is a cloudy Earth in the background, a Soyuz spacecraft anchored to the ISS (bottom), and, with its cargo-bay doors wide open, the Shuttle.

Back jacket: (top row, left) an orbital sunrise viewed from the ISS (see p. 29); (top row, right) as pp. 98–99 below; (middle row, left) a photograph of Jupiter taken by *Voyager 1* (see p. 106); (middle row, center) the red supergiant star V838 Monocerotis, as seen by the *Hubble Space Telescope* (see p. 149); (middle row, right) astronaut Stephen Robinson connected to Canadarm 2 (see p. 42); (bottom row, left) astronaut Soichi Noguchi outside the *Destiny* Laboratory (see p. 40); (bottom row, right) two Soyuz spacecraft (one visible to the camera, the other behind it) docked to the Earth-facing side of the ISS during a crew transfer.

Front endpaper: A startling representation of the thousands of trackable objects—including spacecraft, space junk, and natural debris—that orbit the Earth.

Back endpaper: Light-hearted signs on the hatchway of the ISS's Pressurized Mating Adapter 1.

Page 2: An astronaut grapples with the European Space Agency's *Columbus* Laboratory following its launch aboard the Shuttle *Atlantis* (STS-122) on February 7, 2008.

Pages 4–5: An illustration of the CALIPSO satellite, one of the "A-Train" voyagers in the Earth Observing System group of satellites.

Pages 12–13: Cosmonaut Valery I. Tokarev of the Russian Federal Space Agency waves from the interior of the Pirs Docking Compartment.

Pages 58–59: This image, generated by the *Aqua* satellite, shows variations in the surface temperature of the Earth's oceans (see p. 71).

Pages 98–99: An artist's impression of the European Space Agency's *Venus Express* as it fires its main engine and inserts itself into orbit around Venus (see p. 134).

Pages 140–41: "The Crucible of Creation," located at the center of the Orion Nebula, in an image captured by the *Hubble Space Telescope* (see p. 149).

Pages 176–77: An illustration showing Orion and its crew leaving Earth's atmosphere atop the Ares I launch vehicle (see p. 180).

SPACE STATION
CONSTRUCTION

SPEED
LIMIT
17500